BBC
Bitesize

Bitesize
AQA GCSE (9-1)
MATHEMATICS
REVISION GUIDE
HIGHER

Series Consultant:
Harry Smith

Author:
Navtej Marwaha

Contents

✓ Tick off each topic as you go.

How to use this book

Use the features in this book to focus your revision, track your progress through the topics and practise your exam skills.

② Features to help you revise

Each bite-sized chunk has a **timer** to indicate how long it will take. Use them to plan your revision sessions.

Questions that test **problem-solving** skills are explained in callouts and in the *Problem solving* section at the back.

Test yourself with **exam-style practice** at the end of each page and check your answers at the back of the book.

Scan the **QR codes** to visit the BBC Bitesize website. It will link straight through to more revision resources on that subject.

Complete **worked examples** demonstrate how to approach exam-style questions.

Tick boxes allow you to track the sections you've revised. Revisit each page to embed your knowledge.

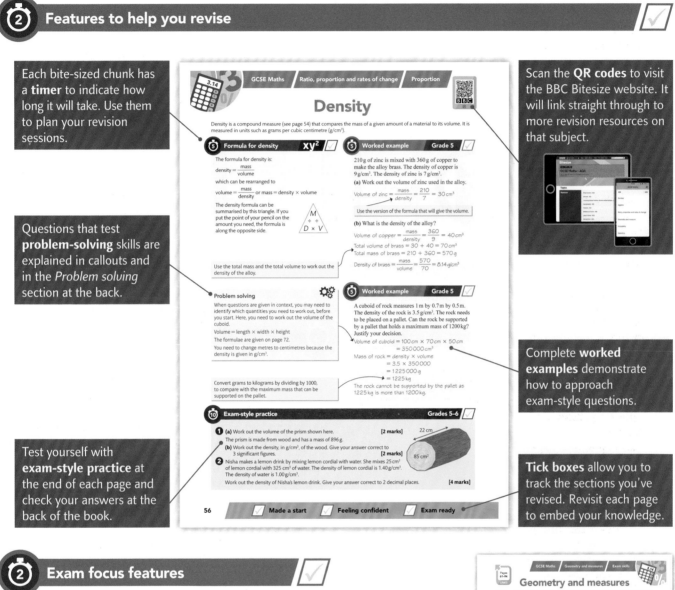

② Exam focus features

The *About your exam* section at the start of the book gives you all the key information about your exams, as well as showing you how to identify the different questions.

Throughout the topic pages you will also find green *Exam skills* pages. These work through an extended exam-style question and provide further opportunities to practise your skills.

② ActiveBook and app

This Revision Guide comes with a **free online edition**. Follow the instructions from inside the front cover to access your ActiveBook.

You can also download the **free BBC Bitesize app** to access revision flash cards and quizzes.

If you do not have a QR code scanner, you can access all the links in this book from your ActiveBook or visit **www.pearsonschools.co.uk/BBCBitesizeLinks**.

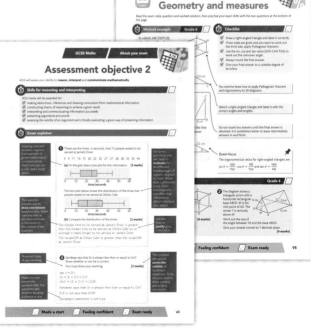

Your Maths GCSE

This page will tell you everything you need to know about the structure of your upcoming AQA Higher GCSE Maths exam.

[5] About the exam papers

You will have to take **three papers** as part of your GCSE Maths qualification.

Paper 1
1 hour 30 minutes
80 marks in total

Paper 2
1 hour 30 minutes
80 marks in total

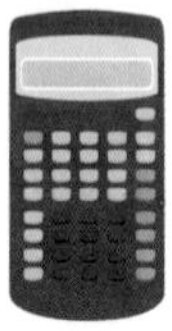

Paper 3
1 hour 30 minutes
80 marks in total

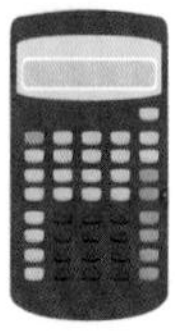

You could be tested on any of the topics you have studied in any of the three written papers. There will be a mixture of question styles on each paper. Papers will usually start with shorter and easier questions and will progress towards harder questions worth more marks at the end of the paper.

[5] Topics

Your AQA GCSE Maths Higher specification is divided into five topics. This pie chart shows the five topics and the proportion of marks that will be allocated to each one:

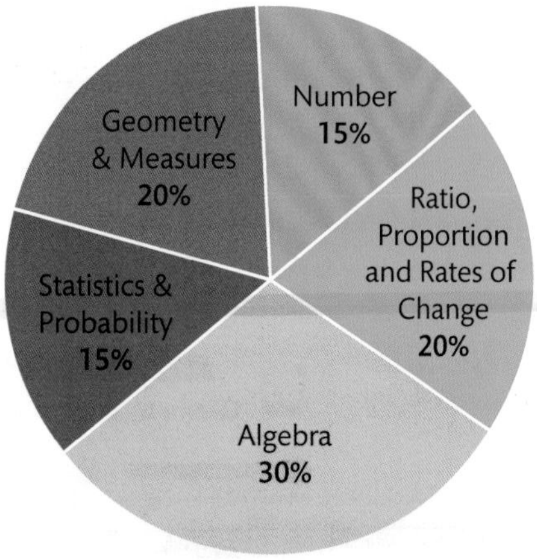
Number 15%
Ratio, Proportion and Rates of Change 20%
Algebra 30%
Statistics & Probability 15%
Geometry & Measures 20%

[5] Assessment objectives

In your exams, marks will be allocated based on three assessment objectives.

Assessment objective 1 (AO1) is about applying and using standard mathematical techniques. About 40% of the marks in your exam will be AO1 marks.

Assessment objective 2 (AO2) is about reasoning, interpreting and communicating mathematically. About 30% of the marks in your exam will be AO2 marks.

Assessment objective 3 (AO3) is about solving unfamiliar problems, and solving problems involving real-life contexts. About 30% of the marks in your exam will be AO3 marks.

You can see some examples of the different assessment objectives on pages vi, vii and viii.

[2] My exam dates

Find out the date and time of each of your GCSE Maths papers and write them in this table.

	Date	AM or PM?
Paper 1		
Paper 2		
Paper 3		

Watch out – no calculators allowed on this paper.

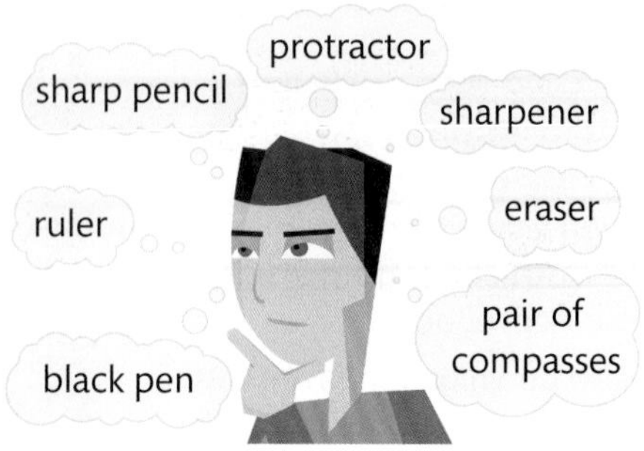
protractor
sharp pencil
sharpener
ruler
eraser
black pen
pair of compasses

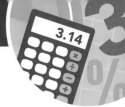

Exam strategies

In your exam you will need to demonstrate your **problem-solving skills**. This page gives some top tips for answering problem-solving questions and how to approach your exam.

5 Problem solving

Lots of questions in your GCSE Maths exam will have a problem-solving element. In your exam, marks will be awarded specifically for:

- following through mathematical processes clearly and correctly
- presenting mathematical proofs
- showing your methods clearly
- solving problems in unfamiliar contexts
- combining techniques in an unfamiliar way.

The **top tip** here is **don't be scared** if a question doesn't look familiar, or if it looks like it requires a lot of steps. You will definitely have covered the techniques needed in your course, so take a deep breath and have a go!

5 Revision advice

- ✓ Make a list of all the topics you need to revise.
- ✓ Create a realistic schedule – you can work backwards from the date of your exam!
- ✓ Start early – don't wait until a few days before your exam!
- ✓ Revise in small chunks and plan to revisit topics again later.
- ✓ Take regular short breaks.
- ✓ Drink plenty of water and eat healthy snacks like fruit or vegetables.
- ✓ Make sure your notes are easy to read – but remember, they don't have to be works of art.
- ✓ Don't work too late at night.
- ✓ Minimise TV and video game time in the run-up to your exams – it will help with concentration.

5 Exam advice

Exam top tips

- Read all the instructions carefully.
- Check that you haven't missed any pages or questions at the end.
- Answer all the questions on each paper.
- Keep explanations short and use correct mathematical language.
- Make sure your answers look sensible.
- Show all your working.
- Read each question carefully before starting to answer it.
- Check your working if you have any spare time at the end.
- Write down some working even if you can't finish a question.
- 1 mark = about 1 minute.
- Write down an answer even if you're not sure it's right.
- Write down all the figures from your calculator display before rounding your answer.

2 Exam language

Sometimes the wording of a question gives you a clue about how to tackle it:

You must show your working.

This means you **have** to show your method and working clearly. If you just write down the correct answer, you might not get the marks.

Explain...

You need to give a **written** answer. Make sure you use the correct mathematical language. You can back your answer up with data or calculations.

You must give a reason for your answer.

Either explain your answer in words, or make sure you have **shown enough working** to justify how you reached the conclusion. Or both!

Assessment objective 1

AO1 will assess your ability to **use and apply standard techniques**.

(5) **Standard mathematical techniques** ✓

AO1 marks will be awarded for:
- ✓ accurately recalling facts, terminology and definitions
- ✓ using and interpreting notation correctly
- ✓ accurately carrying out routine procedures or set tasks requiring multi-step solutions.

(10) **Exam explainer** ✓

Multiplying out brackets is an example of a **routine** task. Questions like this should be familiar from your course and your revision.

Write out $(1 + \sqrt{5})^2$ in the form $a + b\sqrt{5}$. **[3 marks]**

$(1 + \sqrt{5})^2 = (1 + \sqrt{5})(1 + \sqrt{5})$
$= 1 + \sqrt{5} + \sqrt{5} + \sqrt{5}\sqrt{5}$
$= 1 + \sqrt{5} + \sqrt{5} + 5 = 6 + 2\sqrt{5}$
Hence, $a = 6$ and $b = 2$

You can check you have written the answer in the correct form by giving the values of a and b.

1 Given that $A = 2^4 \times 3^3 \times 5$ and $B = 2^3 \times 3^2 \times 5^2$ write down, as a product of powers of prime factors:
(a) the highest common factor (HCF) of A and B **[1 mark]**
(b) the lowest common multiple (LCM) of A and B. **[1 mark]**

'Write down' questions can usually be answered quickly, using familiar techniques. You don't necessarily need to show a lot of working for these questions.

2 Ken rounds a number, x, to one decimal place. The result is 9.4.
Write down the error interval for x. **[2 marks]**

(10) **Exam explainer** ✓

You need to be able to correctly interpret notation like the arrows which show parallel lines on this diagram.

ABCD is a parallelogram.

This is information you should be able to recall.

Prove that triangle ABD is congruent to triangle CDB. **[3 marks]**

BD is common.
$\angle ABD = \angle BDC$ (alternate angles are equal)
$AB = CD$ (opposite sides of a parallelogram)
Triangles ABD and CDB are congruent (SAS).

In harder questions you still need to **recall** standard facts and information, such as angle facts about parallel lines.

Make sure you are confident with standard techniques and processes.

3 Solve these simultaneous equations.
$3x + 5y = 4$
$2x - y = 7$ **[3 marks]**

Your exam will begin with **multiple choice questions**. Read them really carefully. In this question three of the expressions should be equivalent to p^6. You need to circle the one that is not. If you want to change your answer, cross it out then circle the correct answer.

4 Circle the expression that is **not** equivalent to p^6.
$(p^3)^2$ $\dfrac{p^7}{p}$ $\sqrt{p^8}$ $p \times p^5$ **[1 mark]**

✓ **Made a start** ✓ **Feeling confident** ✓ **Exam ready**

Assessment objective 2

AO2 will assess your ability to **reason, interpret** and **communicate mathematically**.

⑤ Skills for reasoning and interpreting ✓

AO2 marks will be awarded for:

- ☑ making deductions, inferences and drawing conclusions from mathematical information
- ☑ constructing chains of reasoning to achieve a given result
- ☑ interpreting and communicating information accurately
- ☑ presenting arguments and proofs
- ☑ assessing the validity of an argument and critically evaluating a given way of presenting information.

⑩ Exam explainer ✓

Drawing neat and accurate diagrams is an example of good mathematical communication. Make sure you use a ruler and a sharp pencil.

This question requires you to **draw conclusions** based on data. Make sure you refer to the data and make a conclusion in the context of the question.

Show each step of your working.

Make sure you answer the question fully. The question asks whether Sandeep is correct or not.

For some questions, you will need to **evaluate** the benefits and disadvantages of a graph, diagram or chart. When data is presented using charts and diagrams you often lose information about individual data values.

Use the data to **justify** your conclusion.

This question asks you to **assess the validity** of Sandeep's statement. To do this you need to show working and write a conclusion.

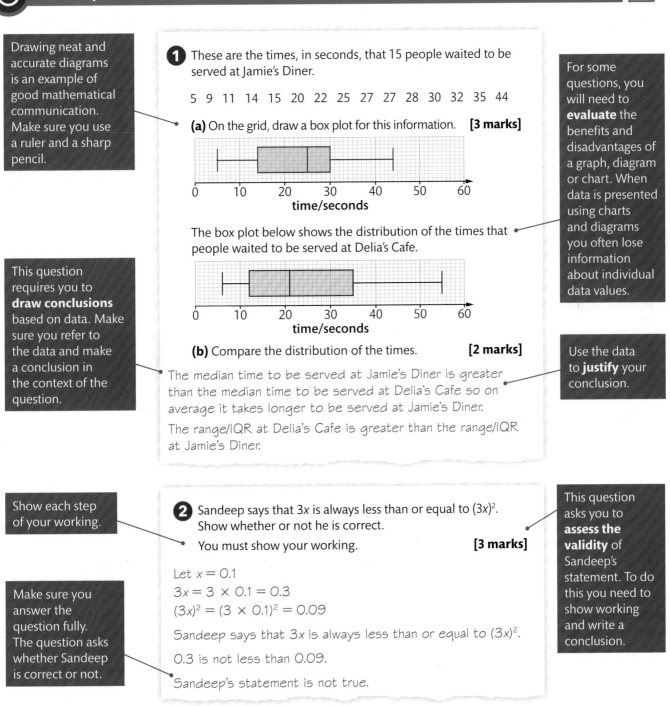

1 These are the times, in seconds, that 15 people waited to be served at Jamie's Diner.

5   9   11   14   15   20   22   25   27   27   28   30   32   35   44

(a) On the grid, draw a box plot for this information.   **[3 marks]**

time/seconds

The box plot below shows the distribution of the times that people waited to be served at Delia's Cafe.

time/seconds

(b) Compare the distribution of the times.   **[2 marks]**

The median time to be served at Jamie's Diner is greater than the median time to be served at Delia's Cafe so on average it takes longer to be served at Jamie's Diner.

The range/IQR at Delia's Cafe is greater than the range/IQR at Jamie's Diner.

2 Sandeep says that $3x$ is always less than or equal to $(3x)^2$.
Show whether or not he is correct.
You must show your working.   **[3 marks]**

Let $x = 0.1$
$3x = 3 \times 0.1 = 0.3$
$(3x)^2 = (3 \times 0.1)^2 = 0.09$

Sandeep says that $3x$ is always less than or equal to $(3x)^2$.

0.3 is not less than 0.09.

Sandeep's statement is not true.

Made a start ✓   Feeling confident ✓   Exam ready ✓   vii

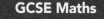

Assessment objective 3

AO3 will assess your ability to **solve problems within mathematics and in other contexts**.

[5] Skills for problem solving

AO3 marks will be awarded for:

- ✓ translating problems in mathematical or non-mathematical contexts into a process or a series of mathematical processes
- ✓ making and using connections between different parts of mathematics
- ✓ interpreting results in the context of the given problem
- ✓ evaluating methods used and results obtained
- ✓ evaluating solutions to identify how they may have been affected by assumptions made.

[10] Exam explainer

A garden is in the shape of a rectangle, ABCD, and a triangle, ABE.

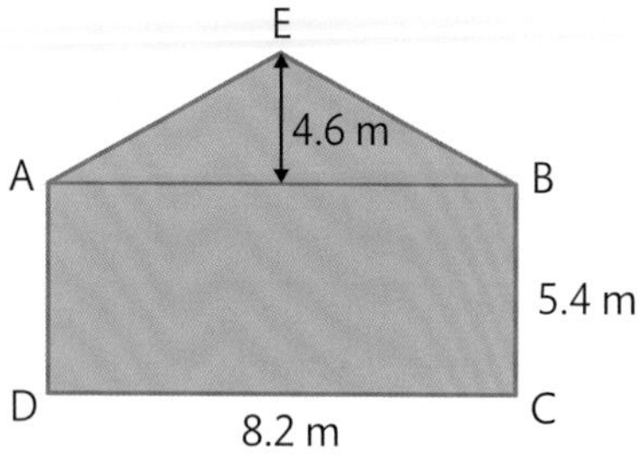

Joan is going to cover the garden with fertiliser.

A bag of fertiliser costs £7.99. One bag of fertiliser will cover an area of 9.4 m^2.

Work out the cost of buying enough fertiliser to cover the garden completely.

[5 marks]

Area of rectangle = $8.2 \times 5.4 = 44.28$ m^2

Area of triangle = $0.5 \times 8.2 \times 4.6 = 18.86$ m^2

Total area of garden = $44.28 + 18.86 = 63.14$ m^2

Number of bags = $63.14 \div 9.4 = 6.71...$

Number of bags needed = 7

Cost of the bags = $7 \times £7.99 = £55.93$

This question is given in a **non-mathematical context**. You need to select the most appropriate mathematical techniques to solve the real-life problem.

Questions that require you to work through a long answer will probably award you marks for using a correct method, so show all of your working.

You need to **evaluate** your solution and give an appropriate answer in context. In real life, you would need to buy a whole number of bags of fertiliser. So round up to the nearest whole number to work out the number of bags needed.

When there are a lot of steps in a question it is a good idea to write down your working at each stage.

Make sure you use the correct units in your answer, and check that it makes sense in the context of the question.

Fractions, decimals and percentages

Fractions, decimals and percentages are different ways of expressing a proportion of a quantity. You should be able to convert between them and use them interchangeably.

2 Converting

- To convert a fraction into a decimal, divide the numerator by the denominator.
  $\frac{3}{5} = 3 \div 5 = 0.6$
- To convert a fraction or a decimal into a percentage, multiply by 100.
  $\frac{5}{8} \times 100 = 62.5\%$ $0.785 \times 100 = 78.5\%$
- To convert a percentage into a fraction, divide by 100.
  $75\% = \frac{75}{100} = \frac{3}{4}$

2 Recurring decimals

$\frac{3}{20} = 0.15$ is a **terminating decimal**.

In a **recurring decimal**, a digit or a group of digits is repeated forever. You can use dots to indicate recurring digits.

$0.\dot{7} = 0.777\,777\,77$

$0.5\dot{7}2\dot{4} = 0.572\,472\,472\,472\,4\ \ldots$

Problem solving

Multiply by 100 because there are two recurring digits. If there is one recurring digit multiply by 10, and if there are three, multiply by 1000.

5 Worked example — Grade 5

Here are four numbers.

Which one is the largest number?

Circle your answer.

[$\frac{2}{3}$] 65% $\frac{3}{5}$ 0.62

0.6666667 0.65 0.6 0.62

Largest is 0.6666667.

Convert each number to a decimal.

5 Worked example — Grade 8

Prove that $0.\dot{3}\dot{9} = \frac{13}{33}$.

$x = 0.393\,939$

$100x = 39.393\,939$

$100x = 39.393\,939$

$x = 0.393\,939$

$99x = 39$

$x = \frac{39}{99} = \frac{13}{33}$

Write the recurring decimal as x. Multiply it by 100 and subtract x to remove the recurring part.

5 Worked example — Grade 4

This pie chart shows how Tony spent his money last month.

How much money did Tony save?

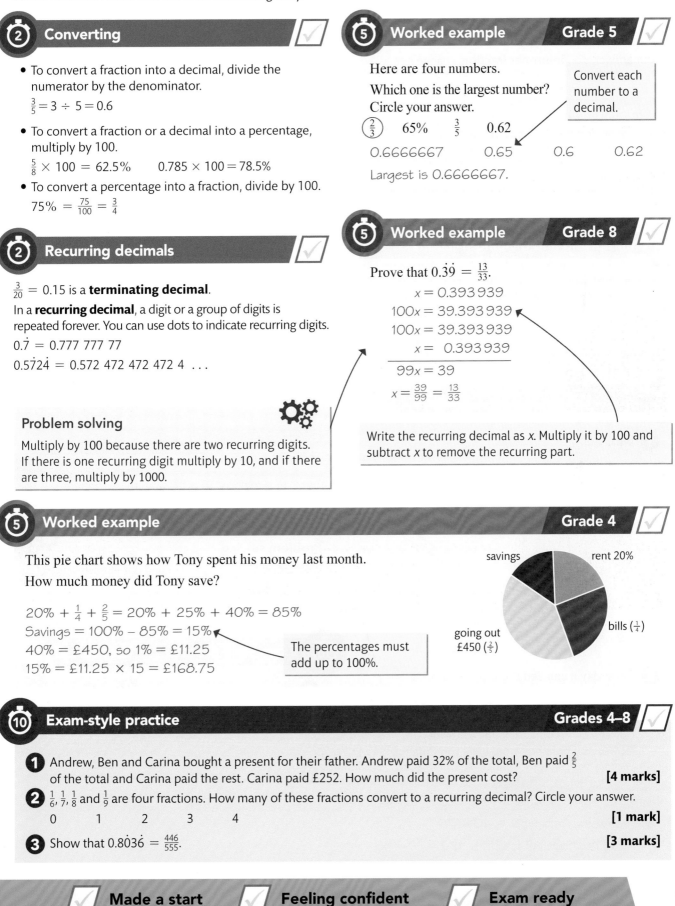

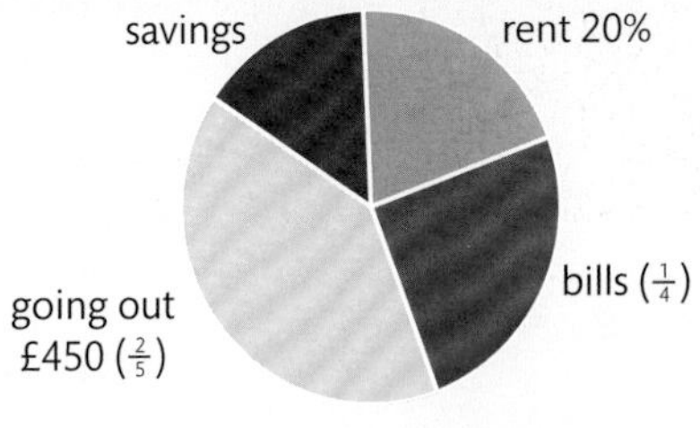
savings
rent 20%
bills (¼)
going out
£450 (⅖)

$20\% + \frac{1}{4} + \frac{2}{5} = 20\% + 25\% + 40\% = 85\%$

Savings $= 100\% - 85\% = 15\%$

$40\% = £450$, so $1\% = £11.25$

$15\% = £11.25 \times 15 = £168.75$

The percentages must add up to 100%.

10 Exam-style practice — Grades 4–8

1. Andrew, Ben and Carina bought a present for their father. Andrew paid 32% of the total, Ben paid $\frac{2}{5}$ of the total and Carina paid the rest. Carina paid £252. How much did the present cost? **[4 marks]**
2. $\frac{1}{6}, \frac{1}{7}, \frac{1}{8}$ and $\frac{1}{9}$ are four fractions. How many of these fractions convert to a recurring decimal? Circle your answer.
   0 1 2 3 4 **[1 mark]**
3. Show that $0.8\dot{0}3\dot{6} = \frac{446}{555}$. **[3 marks]**

Manipulating fractions

You should be able to add, subtract, multiply and divide fractions and mixed numbers without using a calculator.

② Adding and subtracting fractions

Before you can add or subtract fractions you must make sure they have the same denominator. If not, start by finding a common denominator.

If your answer is an **improper fraction**, change it to a mixed number.

$$\frac{3}{4} + \frac{3}{5} = \frac{15}{20} + \frac{12}{20} = \frac{27}{20} = 1\frac{7}{20}$$

When adding or subtracting **mixed numbers** you can either:

- add or subtract the whole number parts and fraction parts separately
- convert both mixed numbers to improper fractions before you add or subtract.

② Worked example — Grade 4

Work out $4\frac{1}{3} - 2\frac{4}{5}$.

$$4\frac{1}{3} - 2\frac{4}{5} = \frac{13}{3} - \frac{14}{5}$$
$$= \frac{65}{15} - \frac{42}{15}$$
$$= \frac{23}{15}$$
$$= 1\frac{8}{15}$$

In this case, $\frac{4}{5} > \frac{1}{3}$ so it will be easier to change the mixed numbers into improper fractions first.

For your final answer, remember to convert the improper fraction to a mixed number.

⑤ Multiplying and dividing fractions

When **multiplying** fractions, multiply the numerators and multiply the denominators. Then cancel if you can.

$$\overset{\div 6}{\frac{3}{4} \times \frac{2}{3} = \frac{6}{12} = \frac{1}{2}}_{\div 6}$$

To **divide** one fraction by another fraction, turn the second fraction upside down and then multiply.

$$\overset{\div 4}{\frac{2}{5} \div \frac{4}{6} = \frac{2}{5} \times \frac{6}{4} = \frac{12}{20} = \frac{3}{5}}_{\div 4}$$

⑩ Worked example — Grade 5

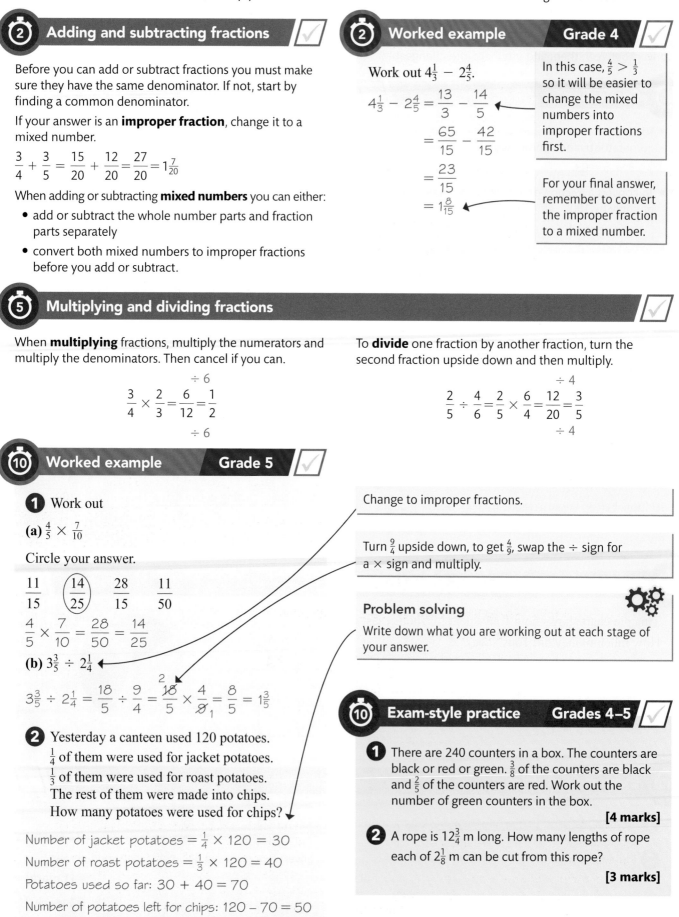

① Work out

(a) $\frac{4}{5} \times \frac{7}{10}$

Circle your answer.

$\frac{11}{15}$ $\boxed{\frac{14}{25}}$ $\frac{28}{15}$ $\frac{11}{50}$

$$\frac{4}{5} \times \frac{7}{10} = \frac{28}{50} = \frac{14}{25}$$

(b) $3\frac{3}{5} \div 2\frac{1}{4}$

$$3\frac{3}{5} \div 2\frac{1}{4} = \frac{18}{5} \div \frac{9}{4} = \frac{\overset{2}{\cancel{18}}}{5} \times \frac{4}{\cancel{9}_1} = \frac{8}{5} = 1\frac{3}{5}$$

② Yesterday a canteen used 120 potatoes.
$\frac{1}{4}$ of them were used for jacket potatoes.
$\frac{1}{3}$ of them were used for roast potatoes.
The rest of them were made into chips.
How many potatoes were used for chips?

Number of jacket potatoes $= \frac{1}{4} \times 120 = 30$

Number of roast potatoes $= \frac{1}{3} \times 120 = 40$

Potatoes used so far: $30 + 40 = 70$

Number of potatoes left for chips: $120 - 70 = 50$

Change to improper fractions.

Turn $\frac{9}{4}$ upside down, to get $\frac{4}{9}$, swap the $\div$ sign for a $\times$ sign and multiply.

Problem solving

Write down what you are working out at each stage of your answer.

⑩ Exam-style practice — Grades 4–5

① There are 240 counters in a box. The counters are black or red or green. $\frac{3}{8}$ of the counters are black and $\frac{2}{5}$ of the counters are red. Work out the number of green counters in the box.

[4 marks]

② A rope is $12\frac{3}{4}$ m long. How many lengths of rope each of $2\frac{1}{8}$ m can be cut from this rope?

[3 marks]

Made a start ☐   Feeling confident ☐   Exam ready ☐

Percentage change

You may be asked to increase or decrease an amount by a percentage of the original value. For instance, in a sale the price of an item is decreased by 10%. There are two methods you can use and you need to know them both.

2 Calculating percentage change

To work out a percentage increase or decrease, you need to write the amount of the increase or decrease as a percentage of the **original amount**. Suppose that, in one year, a plant grew from 80 cm to 92 cm. How can you work out the percentage increase in its height?

Work out the amount of the increase.

92 cm – 80 cm = 12 cm

Write this as a percentage of the original amount.

$\frac{12}{80} \times 100 = 15\%$

5 Worked example — Grade 4

David's weight decreases from 74.5 kg to 69.3 kg.

Work out the percentage decrease in David's weight.

Give your answer correct to 3 significant figures.

Decrease = 74.5 kg – 69.3 kg
= 5.2 kg

Percentage decrease
$= \frac{5.2}{74.5} \times 100$
= 6.98%

Give the decrease as a percentage of the original. The original weight is 74.5 kg.

5 Changing by a percentage

You can increase or decrease an amount by a given percentage either by finding the amount of the change, or by using a multiplier.

How would you work out the total of this bill?

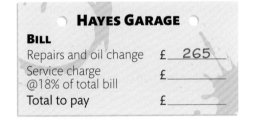

HAYES GARAGE

BILL

Repairs and oil change	£ 265
Service charge @18% of total bill	£
Total to pay	£

Calculating the amount of the change

The amount you need to add on is 18% of £265.

$\frac{18}{100} \times £265 = £47.70$

Add this to the original amount.

£265 + £47.70 = £312.70

Using a multiplier

The total to pay is (100% + 18%) of £265.

118% ÷ 100 = 1.18

The multiplier is 1.18.

£265 × 1.18 = £312.70

To convert the percentage to a decimal divide by 100.

5 Worked example — Grade 4

A set of tyres normally costs £450. In a sale there is a 35% discount.

Work out the sale price of the set of tyres.

$35\% \text{ of } £450 = \frac{35}{100} \times £450$
= £157.50

Work out the amount of the change.

£450 – £157.50 = £292.50

Subtract this from the original amount.

Or use a multiplier. Start by subtracting 35% from 100% and then convert the answer into a decimal.

100% – 35% = 65%
65 ÷ 100 = 0.65
Multiplier is 0.65
£450 × 0.65 = £292.50

You could choose either of these methods in your exam.

10 Exam-style practice — Grade 4

1 A shop sells mobile phones. In January, the shop sold 160 mobile phones. In February, the shop sold 180 mobile phones.

Work out the percentage increase in the number of mobile phones sold from January to February. **[3 marks]**

2 Alan needs to buy some oil for heating at his farm. The capacity of the oil tank is 2300 litres. The tank is already 20% full. Alan wants to fill the tank completely. The price of oil is 73.04 pence per litre. Alan gets 8% off the price of oil. How much does he pay for the oil he needs to buy? **[5 marks]**

Reverse percentages

If you are given an amount after a percentage change, you need to be able to work out the original amount. There are two methods: the **unitary method** and the **multiplier method**.

⏱10 Calculating a reverse percentage ✓

> Amazing reduction!
>
> All prices **reduced by 15%**
>
> **Sale price £204**

You need to calculate a reverse percentage when you are given the final amount **after** a percentage change, and you want to find the original amount.

Using the unitary method

Taking the original price as 100%, after a reduction of 15% the sale price is 100% − 15% = 85%.

The sale price is £204. So 85% is £204.

You need to work out 1%.

1% = £204 ÷ 85 = £2.40

To work out the original price multiply this by 100.

100% = £2.40 × 100 = £240

Using the multiplier method

To work out the multiplier:

100% − 15% = 85%

85% ÷ 100 = 0.85

> Divide by 100 to convert the percentage to a decimal.

The multiplier is 0.85.

So final price = original price × 0.85

To work out the original price, divide the final price by the multiplier.

£204 ÷ 0.85 = £240

⏱5 Worked example | Grade 5 ✓

The price of all plane tickets to Birmingham airport increased by 6%.

(a) The price of a plane ticket from New Delhi to Birmingham increased by £35.28. Work out the price before this increase.

$$6\% = £35.28$$
$$1\% = \frac{£35.28}{6}$$
$$100\% = \frac{£35.28}{6} \times 100 = £588$$

(b) After the increase, the price of a plane ticket from Shanghai to Birmingham was £466.93. Work out the price before this increase.

$$100\% + 6\% = 106\% = 1.06$$
$$£466.93 ÷ 1.06 = £440.50$$

> Part **(a)** is best answered by the unitary method.

> You can use either method to answer part **(b)**.

⏱2 Checklist ✓

✓ To work out the multiplier for an increase, add the percentage increase to 100 and then divide by 100.

$$\text{Multiplier} = \frac{100 + \% \text{ increase}}{100}$$

✓ To work out the multiplier for a decrease, subtract the percentage decrease from 100 and then divide by 100.

$$\text{Multiplier} = \frac{100 - \% \text{ decrease}}{100}$$

⏱10 Exam-style practice | Grade 5 ✓

1 Kim is baking a cake. The cake loses 12% of its mass when it is baked. After the cake is baked its mass is 2.2 kg. Work out the mass of the cake before it is baked. **[3 marks]**

2 Zak and Zoe record their commissions over two years. The table shows their commissions in 2016 and the percentage increase between 2015 and 2016.

Work out whose commission was greater in 2015.

	Commission in 2016	Percentage increase since 2015
Zak	£33 450	4%
Zoe	£34 815	9%

[4 marks]

✓ **Made a start**   ✓ **Feeling confident**   ✓ **Exam ready**

Growth and decay

You can use repeated percentages to model problems involving growth and decay. Typical examples of these are compound interest, population change and depreciation.

⑩ Compound interest

Most bank accounts pay **compound interest**. This means that the amount paid in interest is added to the balance of the account. The next time interest is calculated, the balance will be higher so the amount of interest will be higher. This is an example of **exponential growth**.

Suppose Anjali invests £1600 at 4.2% per annum compound interest. What is the value of Anjali's investment after 4 years?

Using a table of values

$100\% + 4.2\% = 104.2\%$

$104.2 \div 100 = 1.042$ so the multiplier is 1.042.

End of year	Value of investment (£)
1	$1600 \times 1.042 = 1667.20$
2	$1667.20 \times 1.042 = 1737.22$
3	$1737.22 \times 1.042 = 1810.19$
4	$1810.19 \times 1.042 = 1886.21$

Using indices

$1600 \times 1.042 \times 1.042 \times 1.042 \times 1.042 = £1600 \times (1.042)^4$

$£1600 \times (1.042)^4 = £1886.21$

② Depreciation

Something that **depreciates** loses value over time. This is an example of **exponential decay**. Suppose Tina bought a car that cost £15 600. Each year the value of the car depreciates by 16%. What is the value of the car at the end of 3 years?

Using indices

$100\% - 16\% = 84\%$

The multiplier is $84 \div 100 = 0.84$.

The value after 3 years is $£15\,600 \times (0.84)^3 = £9246.18$.

① Revision tips

- ☑ For growth, such as compound interest, the multiplier is more than 1.
- ☑ For decay, such as depreciation, the multiplier is less than 1.

⑩ Worked example — Grade 6

1 Rita needs £55 000 to place a deposit on her new flat. She invests £50 000 at 3.5% compound interest. Does she have enough at the end of three years?

$£50\,000 \times (1.035)^3 = £55\,436$

This is more than £55 000 so she does have enough.

> The multiplier for a 3.5% increase is × 1.035.

2 The value of a car depreciates at the rate of 20% per year. Hanna buys a new car for £36 500. After n years the value of the car is £14 950.40. What is the value of n?

After 2 years: $£36\,500 \times (0.80)^2 = £23\,360$

After 3 years: $£36\,500 \times (0.80)^3 = £18\,688$

After 4 years: $£36\,500 \times (0.80)^4 = £14\,950.40$

The value of n is 4.

> Try different values of n.

⑩ Exam-style practice — Grade 6

1 Sandra invested £3250 in a savings account. She was paid 3.2% per annum compound interest. Work out the number of years it took her to save more than £3650. **[2 marks]**

2 A ball was dropped from a height of 3 m onto horizontal ground. The ball hit the ground and bounced up. Each time the ball bounced, it rose to 65% of its previous height. Work out the height the ball reached after the third bounce. Give your answer correct to 2 decimal places. **[3 marks]**

☑ Made a start   ☑ Feeling confident   ☑ Exam ready

Estimation and counting

You can estimate the answer to a calculation by rounding each number to 1 significant figure and then doing the calculation.

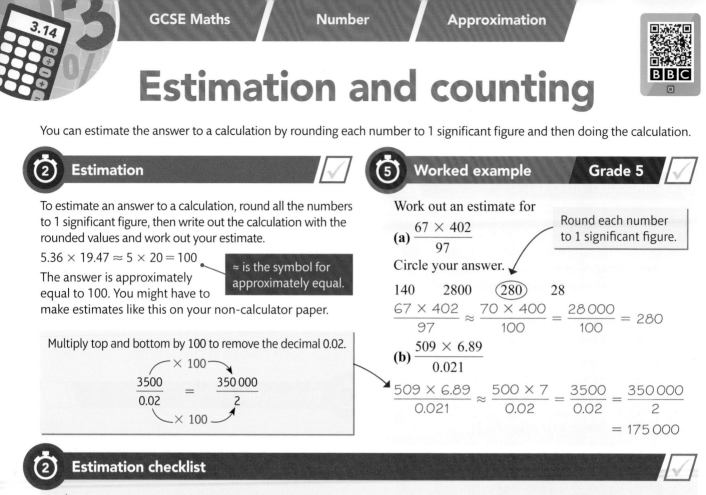

② Estimation

To estimate an answer to a calculation, round all the numbers to 1 significant figure, then write out the calculation with the rounded values and work out your estimate.

$5.36 \times 19.47 \approx 5 \times 20 = 100$

The answer is approximately equal to 100. You might have to make estimates like this on your non-calculator paper.

$\approx$ is the symbol for approximately equal.

Multiply top and bottom by 100 to remove the decimal 0.02.

$$\frac{3500}{0.02} \xrightarrow{\times 100} = \frac{350\,000}{2} \xleftarrow{\times 100}$$

⑤ Worked example — Grade 5

Work out an estimate for

(a) $\dfrac{67 \times 402}{97}$

Round each number to 1 significant figure.

Circle your answer.

140   2800   (280)   28

$$\frac{67 \times 402}{97} \approx \frac{70 \times 400}{100} = \frac{28\,000}{100} = 280$$

(b) $\dfrac{509 \times 6.89}{0.021}$

$$\frac{509 \times 6.89}{0.021} \approx \frac{500 \times 7}{0.02} = \frac{3500}{0.02} = \frac{350\,000}{2}$$
$$= 175\,000$$

② Estimation checklist

- ✓ Always round each number to 1 significant figure.
- ✓ To remove a decimal from the denominator multiply numerator and denominator by 10 or 100 or 1000.
- ✓ Estimate square roots or cube roots by finding nearest square number or cube number.

⑤ Product rule for counting

If there are m ways of choosing one item, and n ways of choosing a second item, then there are a total of $m \times n$ ways of choosing both items. If you have six different t-shirts and four different pairs of jeans, then there are 24 different ways of choosing a t-shirt and a pair of jeans.

$6 \times 4 = 24$

Use the product rule. You could simplify your working using indices: $26^5 \times 10^2$.

⑤ Worked example — Grade 7

The diagram shows a UK number plate. It consists of two letters, followed by two numbers, followed by a further three letters.

MN74 MXU

If the letters can be chosen from any letters in the alphabet and the numbers can be any digit from 0 to 9, work out the total number of combinations that are possible. Give your answer in standard form to 3 significant figures.

$26 \times 26 \times 10 \times 10 \times 26 \times 26 \times 26$
$= 1.19 \times 10^9$

⑩ Exam-style practice — Grades 5–7

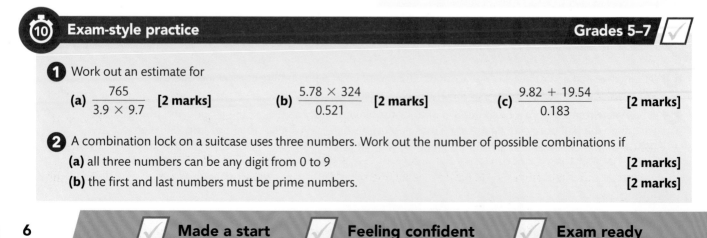

1 Work out an estimate for

(a) $\dfrac{765}{3.9 \times 9.7}$ **[2 marks]**

(b) $\dfrac{5.78 \times 324}{0.521}$ **[2 marks]**

(c) $\dfrac{9.82 + 19.54}{0.183}$ **[2 marks]**

2 A combination lock on a suitcase uses three numbers. Work out the number of possible combinations if

(a) all three numbers can be any digit from 0 to 9 **[2 marks]**

(b) the first and last numbers must be prime numbers. **[2 marks]**

☑ Made a start   ☑ Feeling confident   ☑ Exam ready

Upper and lower bounds

When a quantity is rounded, the actual value could be either higher or lower than the rounded value. You need to be able find an upper bound (UB) and a lower bound (LB) for the actual value.

(5) Rules for upper and lower bounds

An exercise book has a width of 30 cm, rounded to the nearest 10 cm. The actual width could be between 25 and 35 cm.

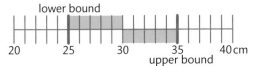

The lower bound for the width is 25 cm and the upper bound is 35 cm.

If you are using upper or lower bounds in a calculation, use these rules to work out the maximum and minimum possible values of the answer.

	Maximum value	Minimum value
Addition	UB + UB	LB + LB
Subtraction	UB − LB	LB − UB
Multiplication	UB × UB	LB × LB
Division	UB ÷ LB	LB ÷ UB

(5) Worked example — Grade 8

The length of a rectangular field is 450 m, to the nearest metre. The width of the field is 243 m, to the nearest metre.

(a) Work out the lower bound for the perimeter of the field.

$$450\,m \qquad\qquad 243\,m$$
$$UB = 450.5\,m \qquad UB = 243.5\,m$$
$$LB = 449.5\,m \qquad LB = 242.5\,m$$

$$LB\ of\ perimeter = 449.5 + 449.5 + 242.5 + 242.5$$
$$= 1384\,m$$

(b) Work out the upper bound for the area of the field.

$$UB\ of\ area = 450.5 \times 243.5$$
$$= 109\,696.75\,m^2$$

> The UB is 0.5 m greater than the rounded measure. The LB is 0.5 m lower.

> Use the rules in the table to find upper and lower bounds in your calculations.

(10) Worked example — Grade 8

1 A ball is thrown vertically upwards with speed v m/s. The height, h metres, to which it rises is given by $h = \dfrac{v^2}{2g}$ where g m/s² is the acceleration due to gravity.

$v = 12.6$ correct to 3 significant figures.

$g = 9.8$ correct to 1 decimal place.

Work out the maximum value of the height, h. Give your answer correct to 3 significant figures.

$$12.6 \qquad\qquad 9.8$$
$$UB = 12.65 \qquad UB = 9.85$$
$$LB = 12.55 \qquad LB = 9.75$$

$$UB\ of\ h = \frac{12.65^2}{2 \times 9.75} = 8.21\,m$$

2 A metal bar has a mass of 1340 g (to the nearest gram) and a volume of 9.4 cm³ (to 1 decimal place).

Given that density $= \dfrac{mass}{volume}$, work out the lower bound for the density of the metal.

Give your answer, in g/cm³, to 3 significant figures.

$$1340\,g \qquad\qquad 9.4\,cm^3$$
$$UB = 1340.5\,g \qquad UB = 9.45\,cm^3$$
$$LB = 1339.5\,g \qquad LB = 9.35\,cm^3$$

$$LB\ of\ density = \frac{mass}{volume} = \frac{1339.5}{9.45} = 142\,g/cm^3$$

> Divide the upper bound by the lower bound to find the maximum possible value of h.

(15) Exam-style practice — Grade 8

1 $I = \dfrac{P}{V}$ where $V = 240$ correct to the nearest 5 and $P = 4200$ correct to the nearest 100. Work out the minimum value of I. Give your answer correct to 3 decimal places. You must show your working. **[3 marks]**

2 Samuel is driving along a French motorway at a constant speed. He has driven 2270 metres, correct to the nearest 10 metres. This took 105 seconds, correct to the nearest 5 seconds. The speed limit for the motorway is 80 km/h. Determine whether Samuel could have exceeded the speed limit. You must show your working. **[4 marks]**

Made a start ☐   Feeling confident ☐   Exam ready ☐

Accuracy and error

BBC

When a question involves upper and lower bounds, you might need to give an answer to an appropriate degree of accuracy.

(5) Appropriate degree of accuracy ✓

To find an appropriate degree of accuracy, start by finding the upper and lower bounds. For an appropriate degree of accuracy, both the upper and lower bounds must round to the **same number**.

Suppose a rectangle is 10.8 cm long, correct to 3 significant figures, and it is 5.37 cm wide, correct to 3 significant figures. By considering bounds, how can you work out the area of the rectangle to a suitable degree of accuracy?

UB of area is $10.85 \times 5.375 = 58.31875$ cm²

LB of area is $10.75 \times 5.365 = 57.67375$ cm²

When the two numbers are rounded they must be equal. Rounded to 2 significant figures:

UB = 58 cm²   and   LB = 58 cm²

So the area is 58 cm² correct to 2 significant figures.

> If you rounded to 3 significant figures you would get:
> UB = 58.3 cm² LB = 57.6 cm²
> These are different, so this is not an appropriate degree of accuracy.

> Use the rules given in the table on page 7 to find upper and lower bounds.

(5) Worked example   Grade 8 ✓

Bo runs 400 metres, correct to the nearest metre. She takes 50.5 seconds, correct to the nearest 0.1 seconds.

(a) Write down the error interval for the distance, d, run by Bo.

$399.5 \leqslant d < 400.5$

(b) Work out the upper and lower bounds of Bo's average speed. Give your answers correct to 5 decimal places.

400	50.5
UB = 400.5	UB = 50.55
LB = 399.5	LB = 50.45

UB speed $= \dfrac{400.5}{50.45} = 7.93855$ m/s

LB speed $= \dfrac{399.5}{50.55} = 7.90307$ m/s

(c) Use your answers to part (a) to write down the value of Bo's average speed to a suitable degree of accuracy. You must explain your answer.

Average speed = 7.9 m/s because the UB and the LB both round to 7.9 m/s when rounded to 2 significant figures.

(5) Worked example   Grade 8 ✓

$t = \dfrac{\sqrt{a}}{b}$

$a = 6.94$ correct to 2 decimal places.

$b = 16.264$ correct to 3 decimal places.

By considering bounds, work out the value of t to a suitable degree of accuracy.

You must show all your working and give a reason for your answer.

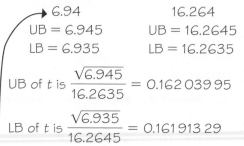

6.94	16.264
UB = 6.945	UB = 16.2645
LB = 6.935	LB = 16.2635

UB of t is $\dfrac{\sqrt{6.945}}{16.2635} = 0.16203995$

LB of t is $\dfrac{\sqrt{6.935}}{16.2645} = 0.16191329$

So $t = 0.162$ to 3 decimal places, because the LB and UB are equal when rounded to that accuracy.

> An **error interval** is a way of writing upper and lower bounds using inequalities. Use $\leqslant$ for the lower bound and $<$ for the upper bound.

(10) Exam-style practice   Grade 8 ✓

1 The time Marcus took to complete a puzzle was 90 seconds.

This time, t, is to the nearest second.

Complete the error interval due to rounding.

____ $\leqslant t <$ ____   **[4 marks]**

2 A lorry can carry a maximum safe load of 30 tonnes to the nearest tonne and delivers pallets of cement to a building site. A pallet of cement weighs 650 kg to the nearest 50 kg.
1 tonne = 1000 kg.

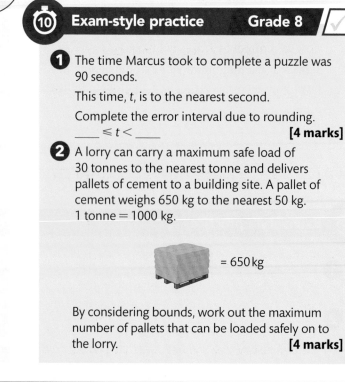
= 650 kg

By considering bounds, work out the maximum number of pallets that can be loaded safely on to the lorry.   **[4 marks]**

☑ **Made a start**   ☑ **Feeling confident**   ☑ **Exam ready**

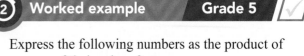 

Factors and primes

Any integer that divides exactly into another given number is a factor of that number. A prime number has only two factors: 1 and itself. A factor that is a prime number is a **prime factor**.

② Prime factors of a number

The numbers 2, 3, 5, 7, 11, 13, 17, 19, 23 and 29 are the first ten **prime numbers**.

If a prime number is a factor of another number it is a **prime factor** of that number.

You can use factor trees to work out prime factors. You keep dividing by prime numbers until all the branches end in a prime factor.

⑤ Lowest common multiple (LCM)

The LCM of two numbers is the lowest number that is a multiple of both numbers.

How can you identify the LCM of 36 and 84?

List their prime factors.

$36 = 2 \times 2 \times 3 \times 3$ and $84 = 2 \times 2 \times 3 \times 7$

For the LCM, compare how many times each factor appears in the two numbers. Look for the maximum number of times that the factor appears in **either** list.

$36 = 2 \times 2 \times 3 \times 3$

$84 = 2 \times 2 \times 3 \times 7$

Both numbers have two 2s, so you need 2×2.

36 has two 3s and 84 has one, so you need 3×3.

84 has one 7 and 36 has none, so you need 7.

Then multiply these together, so the LCM of 36 and 84 is $2 \times 2 \times 3 \times 3 \times 7 = 252$.

⑤ Worked example   Grade 5

(a) Express 120 as the product of its prime factors.

$120 = 2 \times 2 \times 2 \times 3 \times 5$

(b) What is the highest common factor of 120 and 150?

$150 = 2 \times 3 \times 5 \times 5$

$HCF = 2 \times 3 \times 5$

$= 30$

(c) What is the lowest common multiple of 120 and 150?

$LCM = 2 \times 2 \times 2 \times 3 \times 5 \times 5$

$= 600$

Look at all the factors that appear in **either** list. Take the higher number of each factor and multiply the values together. 120 has three 2s where 150 only has one 2, so you use $2 \times 2 \times 2$.

2, 3 and 5 appear in **both** lists.

② Worked example   Grade 5

Express the following numbers as the product of powers of their prime factors.

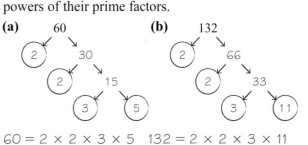

$60 = 2 \times 2 \times 3 \times 5$ $132 = 2 \times 2 \times 3 \times 11$
$60 = 2^2 \times 3 \times 5$ $132 = 2^2 \times 3 \times 11$

You can write the final answer in index form.

⑤ Highest common factor (HCF)

The HCF of two numbers is the highest number that is a factor of both numbers.

To identify the HCF of 36 and 84, list their prime factors.

$36 = 2 \times 2 \times 3 \times 3$ and $84 = 2 \times 2 \times 3 \times 7$

Circle the numbers that appear in **both** lists.

$36 = ②\times②\times③\times 3$ and $84 = ②\times②\times③\times 7$

Both lists include $2 \times 2 \times 3$.

HCF of 36 and 84 is $2 \times 2 \times 3 = 12$.

⑤ Exam-style practice   Grade 5

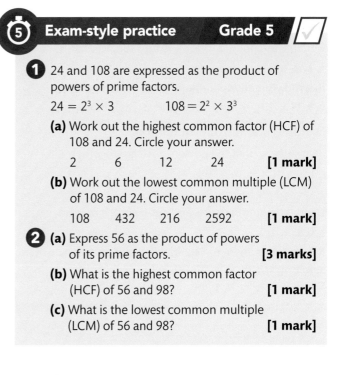

① 24 and 108 are expressed as the product of powers of prime factors.

$24 = 2^3 \times 3$ $108 = 2^2 \times 3^3$

(a) Work out the highest common factor (HCF) of 108 and 24. Circle your answer.

   2   6   12   24   **[1 mark]**

(b) Work out the lowest common multiple (LCM) of 108 and 24. Circle your answer.

   108   432   216   2592   **[1 mark]**

② **(a)** Express 56 as the product of powers of its prime factors.   **[3 marks]**

(b) What is the highest common factor (HCF) of 56 and 98?   **[1 mark]**

(c) What is the lowest common multiple (LCM) of 56 and 98?   **[1 mark]**

Standard form

You can use standard form to write very small and very large numbers in terms of a number between 1 and 10 and a power of 10.

⑤ Numbers in standard form ✓

Notation

A number in standard form is written as:

$A \times 10^n$

where $1 \leqslant A < 10$ and n is an integer.

Standard form is used for writing very small and very large numbers.

$856\,000\,000 = 8.56 \times 10^8$ ⎫ For numbers less
$0.000\,000\,321 = 3.21 \times 10^{-7}$ ⎭ than 1, n is negative.

Writing numbers in standard form

Think of the number as two parts: a number, A, multiplied by a power of 10, 10^n.

$34\,500\,000 = 3.45 \times 10\,000\,000$
$\qquad = 3.45 \times (10 \times 10 \times 10 \times 10 \times 10 \times 10 \times 10)$
$\qquad = 3.45 \times 10^7$

Changing from standard form

To convert the number from standard form, work out the calculation.

$6.34 \times 10^6 = 6.34 \times 1\,000\,000$
$\qquad = 6\,340\,000$

⑤ Multiplying and dividing ✓

Multiplying standard form numbers

Rearrange so that the powers of 10 are together.

$(6.4 \times 10^7) \times (2 \times 10^{-3})$
$= (6.4 \times 2) \times (10^7 \times 10^{-3})$

Multiply the number parts and **add** the powers.

$= 12.8 \times 10^4$

Rewrite your answer in standard form if necessary.

$= 1.28 \times 10^5$

Dividing standard form numbers

Rearrange so that the powers of 10 are together.

$= (3 \times 10^{-8}) \div (6 \times 10^5)$
$= (3 \div 6) \times (10^{-8} \div 10^5)$

Divide the number parts and **subtract** the powers.

$= 0.5 \times 10^{-13}$

Rewrite your answer in standard form if necessary.

$= 5 \times 10^{-14}$

See page 15 for the rules of indices. You need to learn them.

⑤ Worked example — Grade 5 ✓

1 Write these numbers in standard form.

> Divide by a power of 10, to leave a number between 1 and 10.

(a) 562 000

$562\,000 = 5.62 \times 100\,000$
$\qquad\quad = 5.62 \times 10 \times 10 \times 10 \times 10 \times 10$
$\qquad\quad = 5.62 \times 10^5$

(b) 0.000 326

$0.000\,326 = 3.26 \times 0.0001$
$\qquad\quad = 3.26 \times 10^{-4}$

> Take out a factor that is a negative power of 10, to leave a number between 1 and 10.

2 Write 8.29×10^{-5} as an ordinary number.

$8.29 \times 10^{-5} = 8.29 \times 0.00001 = 0.0000829$

> Work out the calculation.

② Use of a calculator ✓

You can enter standard form into a calculator by using the ⌊×10ˣ⌋ button.

Suppose you want to enter 3.2×10^{12} into the calculator.

Enter ⌊3⌋ ⌊.⌋ ⌊2⌋ and then press ⌊×10ˣ⌋ followed by ⌊1⌋ ⌊2⌋.

It will show 3.2×10^{12}.

Then add, subtract, multiply or divide the number as usual.

⑤ Worked example — Grade 5 ✓

Work these out, both using a calculator and without using a calculator.

(a) $(4.5 \times 10^8) \times (3 \times 10^{-6})$

$(4.5 \times 3) \times (10^8 \times 10^{-6}) = 13.5 \times 10^2$
$\qquad\qquad\qquad\qquad\qquad\qquad = 1.35 \times 10^3$

(b) $(3 \times 10^9) \div (4 \times 10^{12})$

$(3 \div 4) \times (10^9 \div 10^{12}) = 0.75 \times 10^{-3} = 7.5 \times 10^{-4}$

⑤ Exam-style practice — Grade 5 ✓

(a) A satellite travels for 5×10^2 hours at a speed of 9×10^4 km/h.

Work out the distance travelled by the satellite. Give your answer in standard form. **[2 marks]**

(b) A second satellite travels 3×10^4 km one month and 4×10^3 km the next month.

Work out the distance travelled by the satellite in the two months. Give your answer in standard form. **[2 marks]**

✓ **Made a start**   ✓ **Feeling confident**   ✓ **Exam ready**

BBC

3.14

Surds

A **rational number** can be expressed in the form $\frac{a}{b}$, where a and b are integers. Any number which cannot be expressed in this form is called an **irrational number**. Square roots of positive integers which are not perfect squares are irrational.

Surds are irrational numbers that contain an irrational square root, such as $\sqrt{2}$, $3 + \sqrt{5}$ or $\frac{7}{\sqrt{11}}$. You can use π and surds to give exact answers.

(5) Simplifying surds

You can use the following rules to simplify surds.

$$\sqrt{ab} = \sqrt{a} \times \sqrt{b} \qquad \sqrt{72} = \sqrt{36} \times \sqrt{2} = 6\sqrt{2}$$

$$\sqrt{\frac{a}{b}} = \frac{\sqrt{a}}{\sqrt{b}} \qquad \sqrt{\frac{5}{9}} = \frac{\sqrt{5}}{\sqrt{9}} = \frac{\sqrt{5}}{3}$$

$$\sqrt{a} \times \sqrt{a} = a$$

A surd is in its **simplest form** when the whole number under the square root sign is the smallest possible number.

$$\sqrt{5} \times \sqrt{5} = 5$$

> Look for a factor of 12 that is a square number.

> There is more about multiplying out brackets on page 13.

(2) Rationalising the denominator

If your answer is a fraction, you might need to **rationalise** the denominator to remove any surd parts. You can do this by multiplying the top and bottom by the surd in the denominator.

$$\frac{21}{\sqrt{3}} \times \frac{\sqrt{3}}{\sqrt{3}} = \frac{21\sqrt{3}}{3} = 7\sqrt{3}$$

(5) Worked example — Grade 7

1 Simplify fully

(a) $\sqrt{12}$

$\sqrt{12} = \sqrt{4 \times 3} = \sqrt{4} \times \sqrt{3} = 2\sqrt{3}$

(b) $5\sqrt{32}$

$5\sqrt{32} = 5\sqrt{16 \times 2} = 5\sqrt{16} \times \sqrt{2}$

$\qquad = 5 \times 4 \times \sqrt{2}$

$\qquad = 20\sqrt{2}$

2 Show that $(5 + \sqrt{7})(5 - \sqrt{7}) = 18$.

$(5 + \sqrt{7})(5 - \sqrt{7})$

$\qquad = 25 + 5\sqrt{7} - 5\sqrt{7} - (\sqrt{7} \times \sqrt{7})$

$\qquad = 25 - 7$

$\qquad = 18$

> Remember that $\sqrt{7} \times \sqrt{7} = 7$

(1) Simplified surds checklist

Make sure:

✓ the numbers under the square roots have no square numbers as factors

✓ you have expanded any brackets

✓ there are no surds in the denominators of fractions.

(5) Worked example — Grade 9

Show that $\dfrac{\sqrt{5}}{\dfrac{1}{\sqrt{5}} + 1}$ can be written as $\dfrac{5\sqrt{5} - 5}{4}$.

$$\frac{\sqrt{5}}{\frac{1}{\sqrt{5}} + 1} = \frac{5}{1 + \sqrt{5}}$$

$$= \frac{5}{1 + \sqrt{5}} \times \frac{1 - \sqrt{5}}{1 - \sqrt{5}}$$

$$= \frac{5 - 5\sqrt{5}}{-4} = \frac{5\sqrt{5} - 5}{4}$$

> Start by multiplying the top and bottom by $\sqrt{5}$ to remove the fraction in the denominator.

Problem solving

If the denominator is in the form $a + \sqrt{b}$, you need to multiply the top and bottom by $a - \sqrt{b}$. So here multiply top and bottom by $1 - \sqrt{5}$.

$(1 + \sqrt{5})(1 - \sqrt{5}) = 1 + 1\sqrt{5} - 1\sqrt{5} - \sqrt{5}\sqrt{5}$

$\qquad = 1 - 5$

$\qquad = -4$

(10) Exam-style practice — Grades 7–9

1 Simplify fully:

(a) $\sqrt{18}$ **(b)** $\sqrt{50}$ **(c)** $\sqrt{45}$ **(d)** $4\sqrt{98}$

[4 marks]

2 Show that $\dfrac{(6 - \sqrt{3})(6 + \sqrt{3})}{\sqrt{11}}$ can be written in the form $k\sqrt{11}$ where k is an integer. **[3 marks]**

✓ **Made a start** ✓ **Feeling confident** ✓ **Exam ready**

Number

Read the exam-style question and worked solution, then practise your exam skills with the two questions at the bottom of the page.

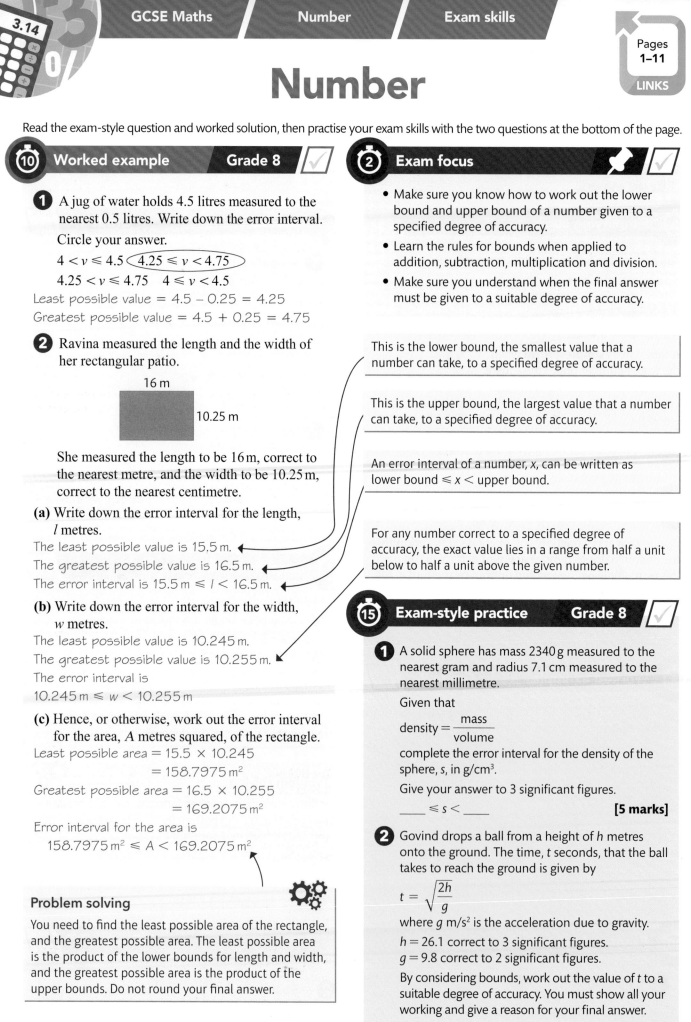

10 Worked example Grade 8 ✓

1 A jug of water holds 4.5 litres measured to the nearest 0.5 litres. Write down the error interval.

Circle your answer.

$4 < v \leqslant 4.5$ ⟨$4.25 \leqslant v < 4.75$⟩

$4.25 < v \leqslant 4.75$ $4 \leqslant v < 4.5$

Least possible value = 4.5 − 0.25 = 4.25
Greatest possible value = 4.5 + 0.25 = 4.75

2 Ravina measured the length and the width of her rectangular patio.

16 m

10.25 m

She measured the length to be 16 m, correct to the nearest metre, and the width to be 10.25 m, correct to the nearest centimetre.

(a) Write down the error interval for the length, l metres.

The least possible value is 15.5 m.
The greatest possible value is 16.5 m.
The error interval is 15.5 m ≤ l < 16.5 m.

(b) Write down the error interval for the width, w metres.

The least possible value is 10.245 m.
The greatest possible value is 10.255 m.
The error interval is
10.245 m ≤ w < 10.255 m

(c) Hence, or otherwise, work out the error interval for the area, A metres squared, of the rectangle.

Least possible area = 15.5 × 10.245
= 158.7975 m²
Greatest possible area = 16.5 × 10.255
= 169.2075 m²
Error interval for the area is
158.7975 m² ≤ A < 169.2075 m²

Problem solving ⚙

You need to find the least possible area of the rectangle, and the greatest possible area. The least possible area is the product of the lower bounds for length and width, and the greatest possible area is the product of the upper bounds. Do not round your final answer.

2 Exam focus 📌 ✓

- Make sure you know how to work out the lower bound and upper bound of a number given to a specified degree of accuracy.
- Learn the rules for bounds when applied to addition, subtraction, multiplication and division.
- Make sure you understand when the final answer must be given to a suitable degree of accuracy.

This is the lower bound, the smallest value that a number can take, to a specified degree of accuracy.

This is the upper bound, the largest value that a number can take, to a specified degree of accuracy.

An error interval of a number, x, can be written as lower bound ≤ x < upper bound.

For any number correct to a specified degree of accuracy, the exact value lies in a range from half a unit below to half a unit above the given number.

15 Exam-style practice Grade 8 ✓

1 A solid sphere has mass 2340 g measured to the nearest gram and radius 7.1 cm measured to the nearest millimetre.

Given that

$$\text{density} = \frac{\text{mass}}{\text{volume}}$$

complete the error interval for the density of the sphere, s, in g/cm³.

Give your answer to 3 significant figures.

____ ≤ s < ____   **[5 marks]**

2 Govind drops a ball from a height of h metres onto the ground. The time, t seconds, that the ball takes to reach the ground is given by

$$t = \sqrt{\frac{2h}{g}}$$

where g m/s² is the acceleration due to gravity.

$h = 26.1$ correct to 3 significant figures.
$g = 9.8$ correct to 2 significant figures.

By considering bounds, work out the value of t to a suitable degree of accuracy. You must show all your working and give a reason for your final answer.

[5 marks]

✓ **Made a start**   ✓ **Feeling confident**   ✓ **Exam ready**

Algebraic expressions

You need to be able to collect like terms, factorise, expand and simplify algebraic expressions. Always give your answers in their simplest form.

② Algebraic expressions

An **algebraic term** is a number or variable, or a combination of numbers and variables, such as 2, x, $4xy$.

An **algebraic expression** is a collection of terms together with arithmetical operators (such as $\times$, $+$).

$2x$, y^2 and $17x - 6y$ are all algebraic expressions.

To **simplify** an expression you need to **collect the like terms**, for example, $3x + 6x = 9x$.

$3x + 6x = 9x$ ◄───── | $3x$ and $6x$ are like terms so can be simplified

| $2x$ and $3y$ are not like terms so this expression cannot be simplified
$2x + 3y$ ◄─────

⑩ Worked example Grade 5

① Expand and simplify

(a) $2(x - 4) + 3(x + 5)$

$= 2x - 8 + 3x + 15$ ◄─────
$= 2x + 3x - 8 + 15$
$= 5x + 7$

(b) $(2x - 3)(x + 4)$

$= 2x(x + 4) - 3(x + 4)$ ◄─────
$= 2x^2 + 8x - 3x - 12$
$= 2x^2 + 5x - 12$

(c) $x(x + 1)(x - 2)$ ◄─────

$= x[x(x - 2) + 1(x - 2)]$
$= x[x^2 - 2x + x - 2]$
$= x[x^2 - x - 2]$
$= x^3 - x^2 - 2x$

② Simplify

(a) $3y + 2x - 4 + 5x + 7$

$= 3y + 2x + 5x - 4 + 7$
$= 3y + 7x + 3$

(b) $5x^2 + 4y^3 + x^2 - 7y^3$

$= 5x^2 + x^2 + 4y^3 - 7y^3$
$= 6x^2 - 3y^3$

③ Factorise

(a) $2x^2 - 4x$

$= 2x(x - 2)$

(b) $12x^3y - 18xy^2$

$= 6xy(2x^2 - 3y)$

(c) $6y^2 - 9xy$

$= 3y(2y - 3x)$

Use BIDMAS:

Brackets
Indices
Division
Multiplication
Addition
Subtraction

② Brackets and factorising

You can **expand brackets** by multiplying the term outside the brackets by everything inside the brackets. The opposite process to expanding brackets is called **factorising**.

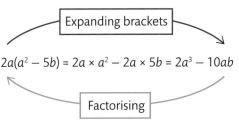

Expanding brackets

$2a(a^2 - 5b) = 2a \times a^2 - 2a \times 5b = 2a^3 - 10ab$

Factorising

You can expand a product of two brackets by multiplying each term in the first bracket by every term in the second bracket:

$(x + 3)(x - 2) = x^2 + 3x - 2x - 6 = x^2 + x - 6$

There are two terms in each bracket so there are four terms in the expanded expression before you simplify.

Multiply out the brackets, then collect like terms.

Multiply everything in the second bracket by everything in the first bracket.

Start by multiplying out $(x + 1)(x - 2)$. Then multiply everything in that expression by x.

Combine the x terms and combine the number terms.

Combine the x^2 terms and combine the y^3 terms.

The HCF of 2 and 4 is 2 and the HCF of x^2 and x is x, so take out a factor of $2x$.

⑮ Exam-style practice Grades 5–6

① Simplify $6y^2 + 7x^3 - 11 + 8x^3 + 17$. **[2 marks]**

② Expand and simplify

(a) $4(y - 3) - 5(3y + 1)$ **[2 marks]**

(b) $x(x + 1)(x - 1)$ **[2 marks]**

③ Factorise fully $20x^2 - 5$. **[2 marks]**

④ Prove algebraically that $(2x + 1)^2 - (2x + 1)$ is an even number for all positive integer values of x.

[3 marks]

BBC

Algebraic formulae

A formula is a mathematical rule. You use algebra to write a formula (the plural of formula is formulae). You should be able to rearrange a formula to **change the subject** and write a formula to describe a simple situation.

---

⑤ Writing a formula ✓

Peter advertises his business in the local magazine.

> **Peter the plumber**
> **Call-out charge £45**
> **£15 per hour**

You can write the total cost as an algebraic formula. Start by defining the variables. For example, you can say the total cost of hiring Peter is T pounds and the number of hours worked is n.

total cost = (number of hours × £15) + call-out charge

$T = 15n + 45$

When you define your variables, you must give their units.

> Use the information given to write y in terms of x. Then use the formula for the volume of a cuboid to find an expression for V in terms of x.

---

⑩ Worked example — Grade 5 ✓

1 Bulbs are sold in packets and in boxes. There are 3 bulbs in a packet. There are 12 bulbs in a box.

Kamran buys x packets of bulbs and y boxes of bulbs.

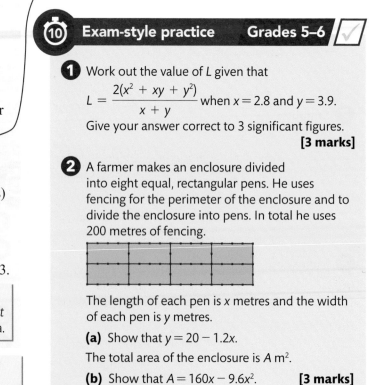

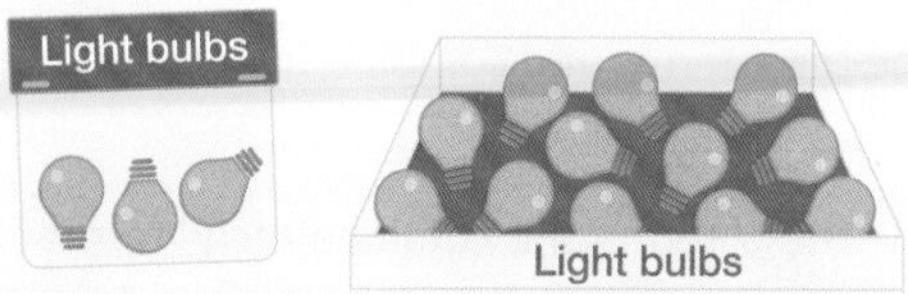

Write down a formula, in terms of x and y, for the total number, N, of bulbs Kamran buys.

$N = 3 \times x + 12 \times y$
$N = 3x + 12y$

2 This formula gives you the distance (s metres) travelled by an object moving with speed (u m/s), in time (t seconds).

$s = ut + 5t^2$

Work out the value of s, when $u = -4$ and $t = 3$.

$s = (-4)(3) + 5(3)^2$
$s = -12 + 45$
$s = 33$

> Substitute the values of u and t into the formula.

> Priority of operations is very important when you are evaluating formulae. Remember to use BIDMAS.

---

⑤ Worked example — Grade 6 ✓

The diagram shows a cuboid of volume V cm³. The length of the cuboid is y cm. The width and height of the cuboid are both x cm.

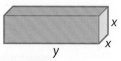

The total length of all the edges of the cuboid is 120 cm and its volume is V cm³.

Show that $V = 30x^2 - 2x^3$.

The sum of all the edges is 120.

$4y + 8x = 120$
$4y = 120 - 8x$
$y = 30 - 2x$

Area of cross-section $= x^2$

$V = x^2y$

> volume of a cuboid = area of cross-section × length

Substitute y into the formula for V.

$V = x^2(30 - 2x)$
$V = 30x^2 - 2x^3$

> The variables are N (the total number of bulbs), x (the number of packets) and y (the number of boxes).

---

⑩ Exam-style practice — Grades 5–6 ✓

1 Work out the value of L given that

$L = \dfrac{2(x^2 + xy + y^2)}{x + y}$ when $x = 2.8$ and $y = 3.9$.

Give your answer correct to 3 significant figures.

[3 marks]

2 A farmer makes an enclosure divided into eight equal, rectangular pens. He uses fencing for the perimeter of the enclosure and to divide the enclosure into pens. In total he uses 200 metres of fencing.

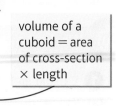

The length of each pen is x metres and the width of each pen is y metres.

(a) Show that $y = 20 - 1.2x$.

The total area of the enclosure is A m².

(b) Show that $A = 160x - 9.6x^2$. **[3 marks]**

✓ **Made a start** ✓ **Feeling confident** ✓ **Exam ready**

Laws of indices

You can use the laws of indices to simplify powers and algebraic expressions.

[5] Basic laws of indices

Learn these laws of indices.

$a^m \times a^n = a^{m+n}$ $\qquad x^4 \times x^6 = x^{4+6} = x^{10}$

$a^m \div a^n = a^{m-n}$ $\qquad x^4 \div x^6 = x^{4-6} = x^{-2}$

$(a^m)^n = a^{mn}$ $\qquad (x^4)^6 = x^{4\times 6} = x^{24}$

$a^0 = 1$ $\qquad x^0 = 1$

[5] Worked example — Grades 5–7

Simplify

(a) $\dfrac{x^5 \times x^7}{x^3}$

$\dfrac{x^5 \times x^7}{x^3} = \dfrac{x^{12}}{x^3} = x^9$

(b) $\left(\dfrac{x^8}{x^5}\right)^2$

$\left(\dfrac{x^8}{x^5}\right)^2 = (x^3)^2 = x^6$

(c) $(2m^4n^2)^5$

$(2m^4n^2)^5 = 2^5 m^{4\times 5} n^{2\times 5} = 32m^{20}n^{10}$

Multiply or divide any numbers first and then use index rules to work out the new powers.

[2] Worked example — Grade 4

Simplify

(a) $p^7 \times p^4$

Use $a^m \times a^n = a^{m+n}$

$p^7 \times p^4 = p^{7+4} = p^{11}$

(b) $p^6 \div p^9$

$p^6 \div p^9 = p^{6-9} = p^{-3}$

(c) $(p^2)^4$

$(p^2)^4 = p^{2\times 4} = p^8$

[1] Indices checklist

- ✓ Only combine powers (indices) when the bases are the same.
- ✓ When you multiply, add the powers.
- ✓ When you divide, subtract the powers.
- ✓ When you raise a power to a power, multiply the powers.

[2] Worked example — Grade 7

Work out the value of n, given that

$\dfrac{y^4 \times y^n}{y} = y^8$

Use the laws of indices to simplify the powers.

$\dfrac{y^{4+n}}{y} = y^8$

Since the base is the same on both sides, the powers must be equal so form an equation and solve it.

$y^{4+n-1} = y^8$

So

$4 + n - 1 = 8$

$n = 5$

[15] Exam-style practice — Grades 4–7

1 Simplify $p^2 \times p^7$. Circle your answer.

p^{14} $\qquad 2p^{14}$ $\qquad 2p^9$ $\qquad p^9$ **[1 mark]**

2 Simplify

(a) $\dfrac{x^4 \times x^6}{x^2}$ **[2 marks]**

(b) $4x^2y^4 \times 3xy$ **[2 marks]**

3 Simplify

(a) $m^8 \div m^2$ **[1 mark]**

(b) $(m^5)^3$ **[1 mark]**

(c) $3w^2y^3 \times 4w^6y$ **[2 marks]**

(d) $\dfrac{32x^6y^8}{4x^2y}$ **[2 marks]**

(e) $(4x^3y^4)^3$ **[2 marks]**

4 Solve to work out the value of n.

$x^6 = \dfrac{x^n \times x^8}{x^5}$ **[3 marks]**

Made a start | Feeling confident | Exam ready

Combining indices

You can use the index laws to simplify expressions involving fractions and negative powers.

5 Further laws of indices

$a^{-n} = \frac{1}{a^n}$ $\qquad 2^{-4} = \frac{1}{2^4} = \frac{1}{16}$

$\left(\frac{a}{b}\right)^n = \frac{a^n}{b^n}$ $\qquad \left(\frac{3}{2}\right)^3 = \frac{3^3}{2^3} = \frac{27}{8}$

$\left(\frac{a}{b}\right)^{-n} = \left(\frac{b}{a}\right)^n$ $\qquad \left(\frac{2}{5}\right)^{-2} = \left(\frac{5}{2}\right)^2 = \frac{5^2}{2^2} = \frac{25}{4}$

$a^{\frac{m}{n}} = \left(a^{\frac{1}{n}}\right)^m$ $\qquad 16^{-\frac{3}{2}} = \left(16^{\frac{1}{2}}\right)^{-3} = 4^{-3} = \frac{1}{4^3} = \frac{1}{64}$

Fractional indices can also be formed as a root: $a^{\frac{m}{n}} = (\sqrt[n]{a})^m$

2 Worked example — Grade 9

1 Solve $4^n = 8^{n-3}$

$(2^2)^n = (2^3)^{n-3}$

$2^{2n} = 2^{3(n-3)}$

$2n = 3(n-3)$

$2n = 3n - 9$

$n = 9$

Remember that $8 = 2^3$, so change the base number to 2.

2 Work out the value of n when $3n^{\frac{3}{2}} = 81^{-\frac{1}{2}}$.

$3n^{\frac{3}{2}} = 81^{-\frac{1}{2}}$

$3n^{\frac{3}{2}} = \frac{1}{81^{\frac{1}{2}}}$

$3n^{\frac{3}{2}} = \frac{1}{9}$

81 = 9^2, so you can find the square root straight away.

$n^{\frac{3}{2}} = \frac{1}{27}$

$n = \left(\frac{1}{27}\right)^{\frac{2}{3}}$

To find n, invert the fraction in the power on the left-hand side and then apply it as a power to the number on the right-hand side.

$n = \left(\frac{1}{27^{\frac{1}{3}}}\right)^2 = \left(\frac{1}{3}\right)^2 = \frac{1}{9}$

1 Indices checklist

- ✓ You can only combine indices if the base numbers are the same.
- ✓ If there is no index, the power is 1, $x = x^1$.
- ✓ Always use brackets to make the working clear, especially when using negative powers.

5 Worked example — Grades 7–8

Work out the value of

(a) 4^{-2} **(b)** $49^{\frac{1}{2}}$

$4^{-2} = \frac{1}{4^2} = \frac{1}{16}$ $\qquad 49^{\frac{1}{2}} = \sqrt{49} = \pm 7$

(c) $27^{-\frac{2}{3}}$

$27^{-\frac{2}{3}} = \left(27^{\frac{1}{3}}\right)^{-2} = 3^{-2} = \frac{1}{3^2} = \frac{1}{9}$

(d) $\left(\frac{81}{16}\right)^{-\frac{3}{4}}$

$\left(\frac{81}{16}\right)^{-\frac{3}{4}} = \left(\frac{16}{81}\right)^{\frac{3}{4}} = \left(\frac{16^{\frac{1}{4}}}{81^{\frac{1}{4}}}\right)^3 = \left(\frac{2}{3}\right)^3 = \frac{2^3}{3^3} = \frac{8}{27}$

2 Solving equations

When solving equations in index form, make sure that the base number is the same in each term.

For example, to work out the value of n when $2^n = 16^{-\frac{1}{2}}$, make the base number on the right-hand side 2.

$2^n = 16^{-\frac{1}{2}}$

$2^n = (2^4)^{-\frac{1}{2}}$

Write 16 as a power of 2. Remember to use brackets.

$2^n = 2^{-2}$

$n = -2$

15 Exam-style practice — Grades 8–9

1 Work out the value of

(a) $8^{\frac{2}{3}}$ **[1 mark]**

(b) $25^{-\frac{1}{2}}$ **[2 marks]**

2 **(a)** Work out the value of $64^{-\frac{1}{2}}$. **[2 marks]**

(b) Write $\frac{\sqrt{3}}{9}$ as 3^n. **[2 marks]**

3 Work out the value of

(a) $(8 \times 4^{-3})^{\frac{1}{3}}$ **[2 marks]**

(b) $\left(\frac{9}{16}\right)^{-\frac{3}{2}}$ **[2 marks]**

4 Solve the equation $9^{n-1} = 27$. **[3 marks]**

5 Work out the value of y for which $2 \times 4^y = 64$. **[4 marks]**

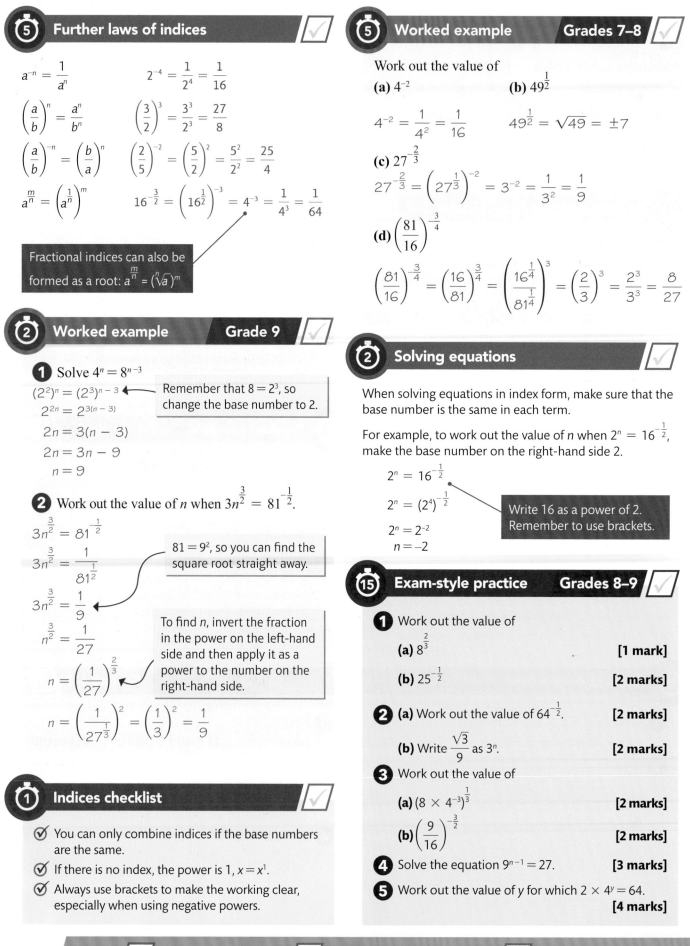

Simple linear equat...

When you solve a linear equation your aim is to find the value of the unknown. You c...
equation and using inverse operations.

(5) Unknown on one side

You may be asked to solve an equation that involves an unknown on one side, such as

$6x + 10 = 34$

You will often need to use **inverse operations**, which are 'opposite' calculations.

The term $+10$ can be removed from the equation by subtracting 10 from each side. This gives:

$6x + 10 - 10 = 34 - 10$
$6x = 24$

To get x by itself, divide both sides by 6.

$\frac{6x}{6} = \frac{24}{6}$
$x = 4$

> Every line of working should include an equals (=) sign. You should align the equals signs underneath each other.

(10) Worked example — Grades 4–5

1 Solve

$5x + 7 = 11$
$5x + 7 - 7 = 11 - 7$
$5x = 4$
$x = \frac{4}{5}$

2 Solve

$7x + 8 = 2x - 3$
$7x - 2x + 8 = 2x - 2x - 3$
$5x + 8 - 8 = -3 - 8$
$5x = -11$
$x = -\frac{11}{5}$

3 Andy, Tom and Chris play hockey. Andy has scored 9 more goals than Chris, Tom has scored 6 more goals than Andy. The total number of goals scored by the three players is 90.

How many goals did they each score?

Andy	Tom	Chris
$x + 9$	$x + 9 + 6$	x

$x + 9 + x + 9 + 6 + x = 90$
$3x + 24 = 90$
$3x = 90 - 24$
$3x = 66$
$x = \frac{66}{3} = 22$

Andy	Tom	Chris
$22 + 9 = 31$	$22 + 9 + 6 = 37$	22

(5) Unknown...

Sometimes an equation has unknow...
To solve such an equation, rearrange it so that a...
unknowns are on one side.

$3 - 4x = 15 - x$
$3 - 4x + 4x = 15 - x + 4x$
$3 = 15 + 3x$
$3 - 15 = 15 - 15 + 3x$
$-12 = 3x$
$\frac{-12}{3} = \frac{3x}{3}$
$-4 = x$ or $x = -4$

> Remove the term $-4x$ from the equation by adding $4x$ to each side.

> To get x by itself, divide both sides by 3.

(2) Equations with brackets

Always multiply out brackets first and then collect like terms. You will be expected to solve an equation such as:

$2(2x + 5) - (3x + 4) = 9(2x + 5)$
$4x + 10 - 3x - 4 = 18x + 45$
$x + 6 = 18x + 45$
$x - x + 6 = 18x - x + 45$
$6 - 45 = 17x + 45 - 45$
$-39 = 17x$
$-\frac{39}{17} = x$
$x = -\frac{39}{17}$

> Multiply out the brackets.

> Collect like terms.

Problem solving

Assign a letter to the unknown value and create an algebraic equation using this letter.

Set up the equation by adding all the expressions and equating them to the total number of goals.

(10) Exam-style practice — Grades 4–5

1 Solve
(a) $6x + 3 = 24$ **[2 marks]**
(b) $3(2x - 1) = 6$ **[2 marks]**
(c) $3x + 7 = 5x - 1$ **[3 marks]**

2 Ann is x years old, Ben is twice as old as Ann. Carl is 4 years younger than Ann. The total of their ages is 92 years.

Work out the age of each person. **[3 marks]**

Linear equations and fractions

...u are dealing with a linear equation that contains fractions, you need to rearrange the equation to remove them, ...ore you can solve it.

⑤ Equations with fractions ✓

You might need to solve an equation such as

$$\frac{x}{5} + \frac{x + 3}{4} = 3$$

You need to remove any fractions before you can solve the equation.

To do this, multiply every term by the lowest common denominator, which is the lowest common multiple (LCM) of the denominators.

$$\frac{20x}{5} + \frac{20(x + 3)}{4} = 3 \times 20$$ •— LCM of 5 and 4 is 20.

$$4x + 5(x + 3) = 60$$ •— Cancel the fractions.

$$4x + 5x + 15 = 60$$
$$9x + 15 = 60$$
$$9x + 15 - 15 = 60 - 15$$
$$9x = 45$$
$$x = 5$$

Sometimes you need to multiply by an **expression** to remove a fraction.

In this case the equation is multiplied by the expression $x + 1$.

⑤ Worked example   Grade 7 ✓

The diagram shows the length, in centimetres, of each side of a rectangle. The perimeter of the rectangle is P cm.

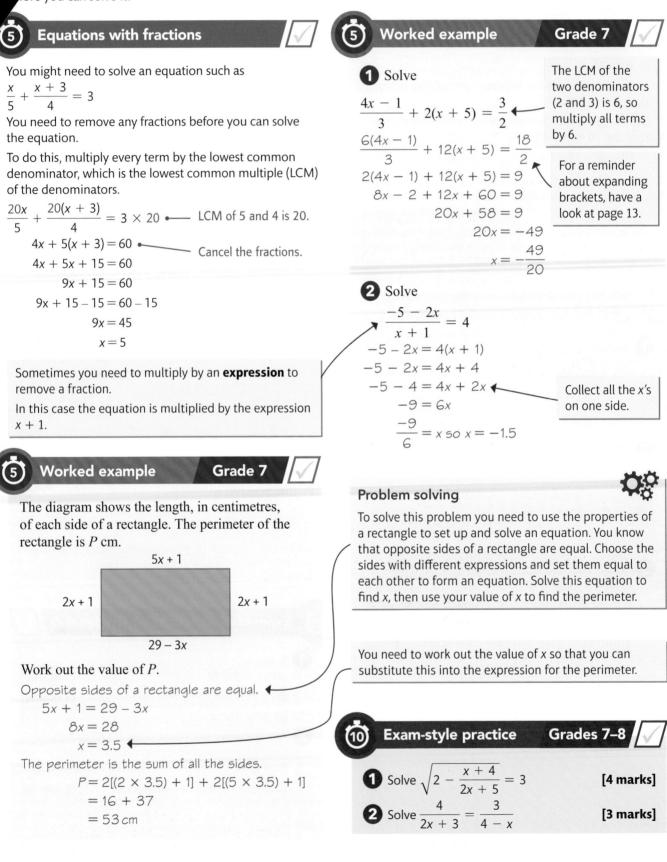

5x + 1

2x + 1   2x + 1

29 − 3x

Work out the value of P.

Opposite sides of a rectangle are equal. ◄

$$5x + 1 = 29 - 3x$$
$$8x = 28$$
$$x = 3.5$$ ◄

The perimeter is the sum of all the sides.

$$P = 2[(2 \times 3.5) + 1] + 2[(5 \times 3.5) + 1]$$
$$= 16 + 37$$
$$= 53 \text{ cm}$$

⑤ Worked example   Grade 7 ✓

1 Solve

$$\frac{4x - 1}{3} + 2(x + 5) = \frac{3}{2}$$ ◄

$$\frac{6(4x - 1)}{3} + 12(x + 5) = \frac{18}{2}$$

$$2(4x - 1) + 12(x + 5) = 9$$
$$8x - 2 + 12x + 60 = 9$$
$$20x + 58 = 9$$
$$20x = -49$$
$$x = -\frac{49}{20}$$

> The LCM of the two denominators (2 and 3) is 6, so multiply all terms by 6.

> For a reminder about expanding brackets, have a look at page 13.

2 Solve

$$\frac{-5 - 2x}{x + 1} = 4$$

$$-5 - 2x = 4(x + 1)$$
$$-5 - 2x = 4x + 4$$
$$-5 - 4 = 4x + 2x$$ ◄
$$-9 = 6x$$
$$\frac{-9}{6} = x \text{ so } x = -1.5$$

> Collect all the x's on one side.

Problem solving ⚙

To solve this problem you need to use the properties of a rectangle to set up and solve an equation. You know that opposite sides of a rectangle are equal. Choose the sides with different expressions and set them equal to each other to form an equation. Solve this equation to find x, then use your value of x to find the perimeter.

You need to work out the value of x so that you can substitute this into the expression for the perimeter.

⑩ Exam-style practice   Grades 7–8 ✓

1 Solve $\sqrt{2 - \dfrac{x + 4}{2x + 5}} = 3$ [4 marks]

2 Solve $\dfrac{4}{2x + 3} = \dfrac{3}{4 - x}$ [3 marks]

Simultaneous equations

Simultaneous equations are equations that are written in terms of the same two unknowns. In this section, you will review solving two simultaneous linear equations.

5 Elimination method

Here are a pair of simultaneous equations

$3x + 4y = 5$ (1)

$2x - 3y = 9$ (2)

Start by numbering each equation, so you can refer to them in your working.

If necessary, multiply one or both of the equations so that the **coefficients** of one unknown are the same.

$6x + 8y = 10$ (1) × 2

$6x - 9y = 27$ (2) × 3

Subtract equation (2) from equation (1), to eliminate the *x* terms.

$17y = -17$

$y = -1$

Now substitute this unknown into one of the original equations to find the other unknown.

$3x + 4(-1) = 5$

$3x = 9$

$x = 3$

Check the answer by substituting the unknown into the other (unused) original equation.

$2x - 3y = 9$

$2(3) - 3(-1) = 9$

$9 = 9$

Problem solving

You need to convert the word problem into a pair of simultaneous equations. Let the weight of one apple be *a* grams, and the weight of one pear be *p* grams.

2 Graphical method

You can solve simultaneous linear equations by drawing up a table of values and plotting a graph. For example, consider $2x + 3y = 18$ and $3x - 4y = 6$.

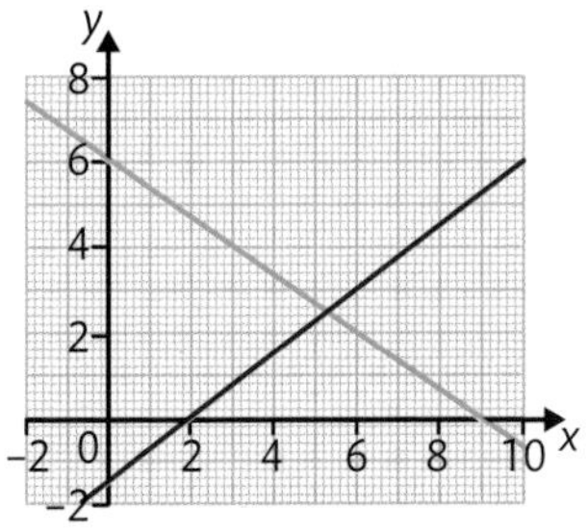

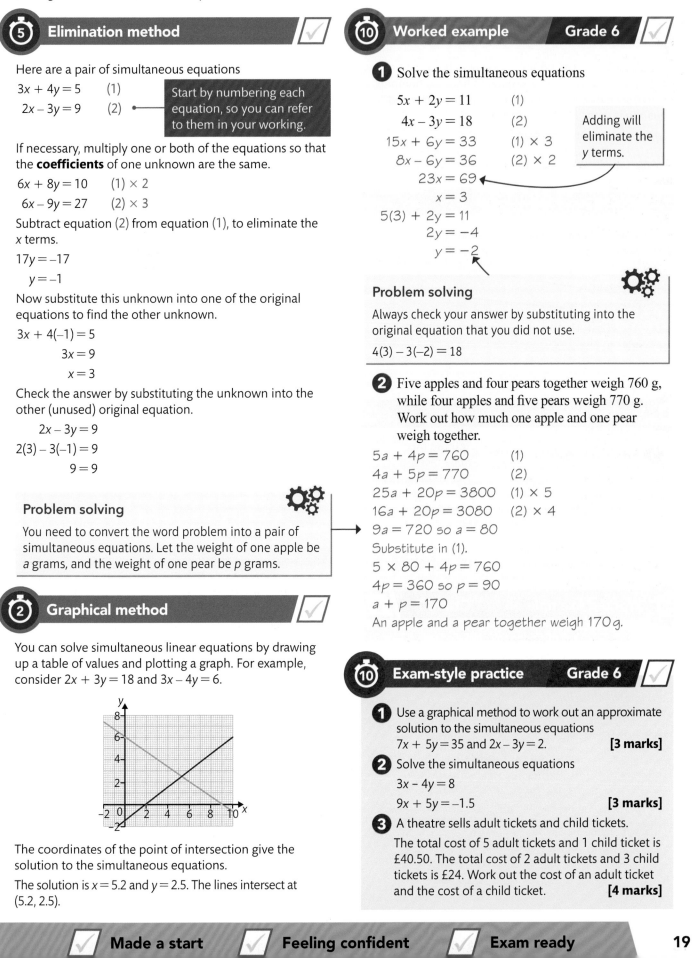

The coordinates of the point of intersection give the solution to the simultaneous equations.

The solution is $x = 5.2$ and $y = 2.5$. The lines intersect at (5.2, 2.5).

10 Worked example — Grade 6

1 Solve the simultaneous equations

$5x + 2y = 11$ (1)

$4x - 3y = 18$ (2)

$15x + 6y = 33$ (1) × 3

$8x - 6y = 36$ (2) × 2

$23x = 69$

$x = 3$

$5(3) + 2y = 11$

$2y = -4$

$y = -2$

Adding will eliminate the *y* terms.

Problem solving

Always check your answer by substituting into the original equation that you did not use.

$4(3) - 3(-2) = 18$

2 Five apples and four pears together weigh 760 g, while four apples and five pears weigh 770 g. Work out how much one apple and one pear weigh together.

$5a + 4p = 760$ (1)

$4a + 5p = 770$ (2)

$25a + 20p = 3800$ (1) × 5

$16a + 20p = 3080$ (2) × 4

$9a = 720$ so $a = 80$

Substitute in (1).

$5 \times 80 + 4p = 760$

$4p = 360$ so $p = 90$

$a + p = 170$

An apple and a pear together weigh 170 g.

10 Exam-style practice — Grade 6

1 Use a graphical method to work out an approximate solution to the simultaneous equations $7x + 5y = 35$ and $2x - 3y = 2$. **[3 marks]**

2 Solve the simultaneous equations

$3x - 4y = 8$

$9x + 5y = -1.5$ **[3 marks]**

3 A theatre sells adult tickets and child tickets.

The total cost of 5 adult tickets and 1 child ticket is £40.50. The total cost of 2 adult tickets and 3 child tickets is £24. Work out the cost of an adult ticket and the cost of a child ticket. **[4 marks]**

Quadratic equations

Quadratic equations have the form $ax^2 + bx + c = 0$, $a \neq 0$. You will need to know how to solve a quadratic equation by factorisation, without using a calculator, by completing the square or by the quadratic formula.

2 Solving $x^2 + bx + c = 0$

To factorise the left-hand side of a quadratic equation in the form $x^2 + bx + c = 0$, you need to find two numbers p and q such that $pq = c$ and $p + q = b$. The factorised equation is then $(x + p)(x + q) = 0$.

So the solutions are $x = -p$ and $x = -q$.

2 Worked example — Grade 5

Solve $x^2 - x - 72 = 0$

$(x + 8)(x - 9) = 0$

$x + 8 = 0$ or $x - 9 = 0$

$x = -8$ or $x = 9$

8 and −9 are a factor pair of −72 that add up to −1.

You can check your factorisation by expanding the brackets.

5 Solving $ax^2 + bx + c = 0$

You can solve quadratic equations of the form $ax^2 + bx + c = 0$ by finding values p and q such that $p + q = b$ and $pq = ac$.

$3x^2 - 7x + 2 = 0$ — $b = -7$ and $ac = 6$ so $p = -1$ and $q = -6$ are suitable values.

$3x^2 - x - 6x + 2 = 0$ — Use your values of p and q to write $-7x$ as $-x - 6x$.

$x(3x - 1) - 2(3x - 1) = 0$ — Factorise the first pair of terms and the last pair of terms.

$(3x - 1)(x - 2) = 0$ — Factorise again then set each factor equal to 0 to solve $x = \frac{1}{3}$ or $x = 2$.

5 Worked example — Grade 6

1 Solve $3x^2 - 5x = 0$.

$3x^2 - 5x = 0$

$x(3x - 5) = 0$

$x = 0$ or $3x - 5 = 0$

$x = 0$ or $x = \frac{5}{3}$

You can take out a factor of x from both terms. Remember there are still two solutions. $x = 0$ or $x = \frac{5}{3}$.

2 Solve $4x^2 - 9 = 0$.

$4x^2 - 9 = 0$

$(2x - 3)(2x + 3) = 0$

$2x - 3 = 0$ or $2x + 3 = 0$

$x = \frac{3}{2}$ or $x = -\frac{3}{2}$

This is the **difference of two squares**. Use $x^2 - a^2 = (x - a)(x + a)$.

5 Worked example — Grade 7

The diagram shows a trapezium. All measurements are in centimetres. The area of the trapezium is 65 cm².

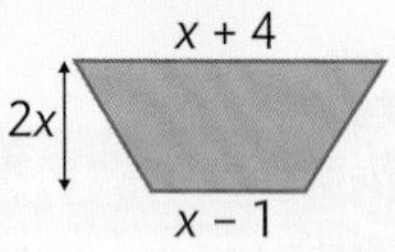

x + 4
2x
x − 1

Work out the value of x.

Area of trapezium $= \frac{1}{2}(x + 4 + x - 1)(2x)$

$= \frac{1}{2}(2x + 3)(2x)$

$= 2x^2 + 3x$

Hence: $2x^2 + 3x = 65$

$2x^2 + 3x - 65 = 0$

$(2x + 13)(x - 5) = 0$

$2x + 13 = 0$ or $x - 5 = 0$

$x = -\frac{13}{2}$ or $x = 5$

$x = 5$ because it cannot be negative.

Always write the equation in the form $ax^2 + bx + c = 0$.

10 Exam-style practice — Grade 7

1 Solve

(a) $9x^2 - 25 = 0$ **[2 marks]**

(b) $3x^2 - 11x = 4$ **[3 marks]**

2 The diagram shows a trapezium. All measurements are in centimetres. The area of the trapezium is 55 cm².

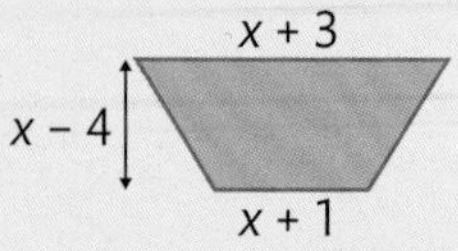

x + 3
x − 4
x + 1

(a) Show that $x^2 - 2x - 63 = 0$. **[2 marks]**

(b) Work out the value of x. **[2 marks]**

Mixed simultaneous equations

You may be asked to solve simultaneous equations that involve a quadratic equation and a linear equation. You should solve simultaneous equations like this using **substitution**.

② Substitution checklist

- ☑ Rearrange the linear equation to $x = ...$ or $y = ...$
- ☑ **Substitute** into the quadratic equation.
- ☑ Rewrite the resulting equation in the form $ax^2 + bx + c = 0$ and solve equation.
- ☑ Substitute each x-value into the linear equation to find the corresponding y-value.
- ☑ Write your final answer as coordinates of points.
- ☑ If the quadratic equation does not factorise you can either use the formula or complete the square.

You could substitute x or y into the quadratic equation. It is easier to substitute for y because there will be no fractions.

If the quadratic equation does not factorise you can either use the quadratic formula or complete the square. Have a look at pages 22 and 23.

⑤ Graphical method

You can solve simultaneous equations by drawing a graph. For example, to solve $x^2 + y^2 = 25$ and $y - x = 1$, rewrite the equations so that you can draw up tables of values. Then the graph looks like this.

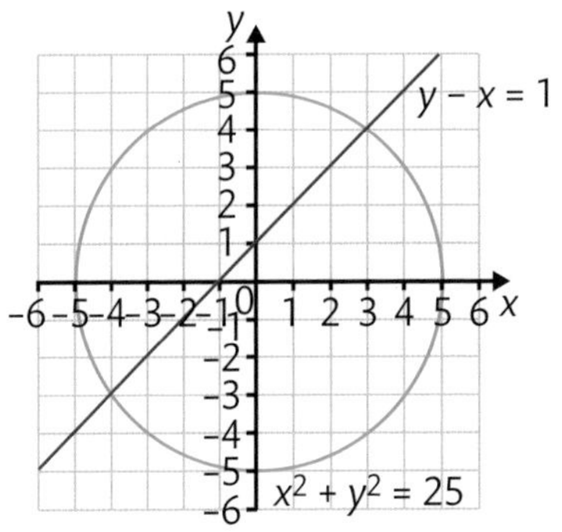

The coordinates of the points of intersection give the solution to the simultaneous equations.

The solutions are $x = -4$, $y = -3$ and $x = 3$, $y = 4$, giving the points $(-4, -3)$ and $(3, 4)$.

⑩ Worked example — Grade 8

① Solve the simultaneous equations:

$2x - y = 7$

$x^2 + y^2 = 34$

You must show your working.

Do not use trial and improvement.

$$y = 2x - 7$$
$$x^2 + (2x - 7)^2 = 34$$ ← Always substitute from the linear equation into the quadratic equation.
$$x^2 + 4x^2 - 28x + 49 = 34$$
$$5x^2 - 28x + 15 = 0$$
$$(5x - 3)(x - 5) = 0$$
$$x = \tfrac{3}{5} \text{ and } x = 5$$
$$y = 2\left(\tfrac{3}{5}\right) - 7 = -\tfrac{29}{5} \text{ and } y = 2(5) - 7 = 3$$

Solutions are $x = \tfrac{3}{5}$, $y = -\tfrac{29}{5}$ and $x = 5$, $y = 3$ or $\left(\tfrac{3}{5}, -\tfrac{29}{5}\right)$ and $(5, 3)$.

② Solve the simultaneous equations:

$y = 2x^2 + 3x - 2$

$y = 2x - 1$

You must show your working.

Do not use trial and improvement.

$$2x^2 + 3x - 2 = 2x - 1$$ ← As both equations have y as the subject, set the RHSs equal.
$$2x^2 + x - 1 = 0$$
$$(2x - 1)(x + 1) = 0$$
$$x = \tfrac{1}{2} \text{ and } x = -1$$

So $y = 2\left(\tfrac{1}{2}\right) - 1 = 0$ and $y = 2(-1) - 1 = -3$

Solutions are $x = \tfrac{1}{2}$, $y = 0$ and $x = -1$, $y = -3$ or $\left(\tfrac{1}{2}, 0\right)$ and $(-1, -3)$.

⑮ Exam-style practice — Grade 8

Solve the following simultaneous equations.

(a) $y = 20 - 3x$

$y = 2x^2$ [6 marks]

(b) $x^2 + y^2 = 36$

$x = 2y + 6$ [6 marks]

(c) $x^2 + y^2 = 9$

$x + y = 2$ [6 marks]

Give your answers correct to 2 decimal places.

You must show your working.

Do not use trial and improvement.

☑ **Made a start**   ☑ **Feeling confident**   ☑ **Exam ready**

Completing the square

Some quadratic equations cannot be factorised. To solve them, you **complete the square**.

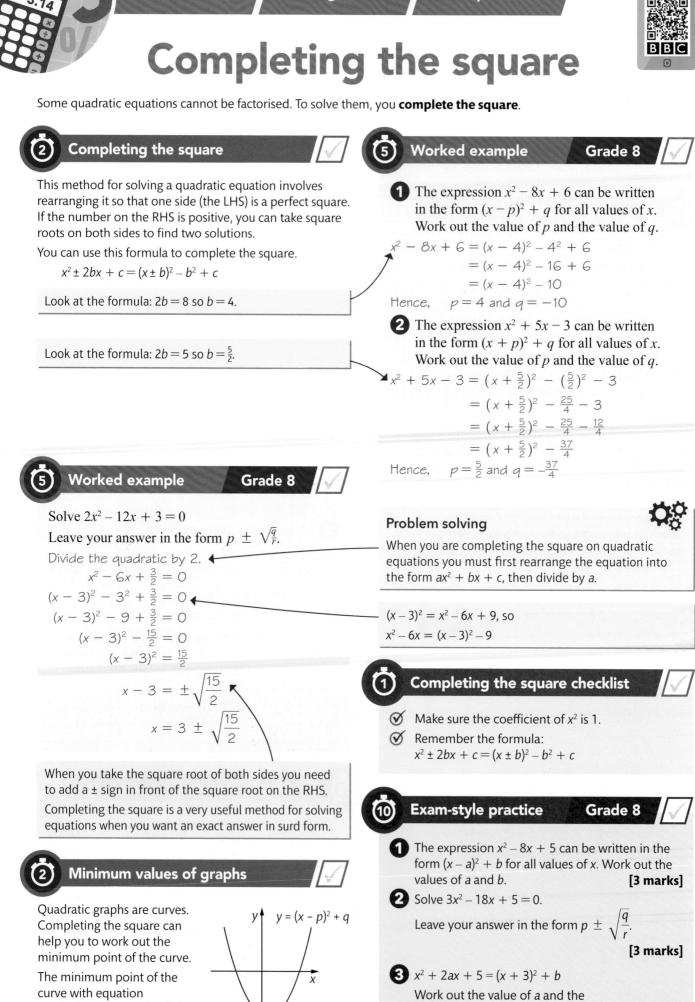

② Completing the square

This method for solving a quadratic equation involves rearranging it so that one side (the LHS) is a perfect square. If the number on the RHS is positive, you can take square roots on both sides to find two solutions.

You can use this formula to complete the square.

$$x^2 \pm 2bx + c = (x \pm b)^2 - b^2 + c$$

Look at the formula: $2b = 8$ so $b = 4$.

Look at the formula: $2b = 5$ so $b = \frac{5}{2}$.

⑤ Worked example — Grade 8

1 The expression $x^2 - 8x + 6$ can be written in the form $(x - p)^2 + q$ for all values of x. Work out the value of p and the value of q.

$$x^2 - 8x + 6 = (x - 4)^2 - 4^2 + 6$$
$$= (x - 4)^2 - 16 + 6$$
$$= (x - 4)^2 - 10$$

Hence, $p = 4$ and $q = -10$

2 The expression $x^2 + 5x - 3$ can be written in the form $(x + p)^2 + q$ for all values of x. Work out the value of p and the value of q.

$$x^2 + 5x - 3 = \left(x + \frac{5}{2}\right)^2 - \left(\frac{5}{2}\right)^2 - 3$$
$$= \left(x + \frac{5}{2}\right)^2 - \frac{25}{4} - 3$$
$$= \left(x + \frac{5}{2}\right)^2 - \frac{25}{4} - \frac{12}{4}$$
$$= \left(x + \frac{5}{2}\right)^2 - \frac{37}{4}$$

Hence, $p = \frac{5}{2}$ and $q = -\frac{37}{4}$

⑤ Worked example — Grade 8

Solve $2x^2 - 12x + 3 = 0$

Leave your answer in the form $p \pm \sqrt{\frac{q}{r}}$.

Divide the quadratic by 2.
$$x^2 - 6x + \frac{3}{2} = 0$$
$$(x - 3)^2 - 3^2 + \frac{3}{2} = 0$$
$$(x - 3)^2 - 9 + \frac{3}{2} = 0$$
$$(x - 3)^2 - \frac{15}{2} = 0$$
$$(x - 3)^2 = \frac{15}{2}$$

$$x - 3 = \pm\sqrt{\frac{15}{2}}$$

$$x = 3 \pm \sqrt{\frac{15}{2}}$$

When you take the square root of both sides you need to add a ± sign in front of the square root on the RHS.

Completing the square is a very useful method for solving equations when you want an exact answer in surd form.

Problem solving

When you are completing the square on quadratic equations you must first rearrange the equation into the form $ax^2 + bx + c$, then divide by a.

$(x - 3)^2 = x^2 - 6x + 9$, so
$x^2 - 6x = (x - 3)^2 - 9$

① Completing the square checklist

☑ Make sure the coefficient of x^2 is 1.

☑ Remember the formula:
$x^2 \pm 2bx + c = (x \pm b)^2 - b^2 + c$

② Minimum values of graphs

Quadratic graphs are curves. Completing the square can help you to work out the minimum point of the curve.

The minimum point of the curve with equation $y = (x - p)^2 + q$ is at (p, q).

$y = (x - p)^2 + q$

⑩ Exam-style practice — Grade 8

1 The expression $x^2 - 8x + 5$ can be written in the form $(x - a)^2 + b$ for all values of x. Work out the values of a and b. **[3 marks]**

2 Solve $3x^2 - 18x + 5 = 0$.

Leave your answer in the form $p \pm \sqrt{\frac{q}{r}}$. **[3 marks]**

3 $x^2 + 2ax + 5 = (x + 3)^2 + b$

Work out the value of a and the value of b. **[3 marks]**

☑ Made a start ☑ Feeling confident ☑ Exam ready

The quadratic formula

You can use the quadratic formula to solve quadratic equations.

5 The quadratic formula xy^2

When you cannot factorise a quadratic equation or complete the square, you can use the **quadratic formula**. The solutions of the quadratic equation $ax^2 + bx + c = 0$ where $a \neq 0$ are given by

$$x = \frac{-b \pm \sqrt{b^2 - 4ac}}{2a}$$

To solve the equation, substitute the values of a, b and c into the formula.

You need to remember this formula for your exams.

2 Quadratic formula checklist

- Write out the quadratic equation in the form $ax^2 + bx + c = 0$.
- Write down the values of a, b and c.
- Show your substitution into the formula clearly.
- Use brackets when using negative numbers.
- The symbol ± means that you need to do two calculations.

5 Worked example Grade 7

Solve $3x^2 + 7x - 13 = 0$.

Give your answers correct to 3 significant figures.

$a = 3 \quad b = 7 \quad c = -13$

$$x = \frac{-7 \pm \sqrt{7^2 - (4 \times 3 \times -13)}}{2 \times 3}$$

$$x = \frac{-7 \pm \sqrt{205}}{6}$$

$$x = \frac{-7 - \sqrt{205}}{6} \quad \text{or} \quad x = \frac{-7 + \sqrt{205}}{6}$$

$x = -3.55$ or $x = 1.22$

The quadratic equation does not factorise, so you need to use the quadratic formula.

If the question asks you to find solutions to a given degree of accuracy it is a clue that you should use the quadratic formula.

The graph of $y = 3x^2 + 7x - 13$ cuts the x-axis at $x = -3.55$ and $x = 1.22$.

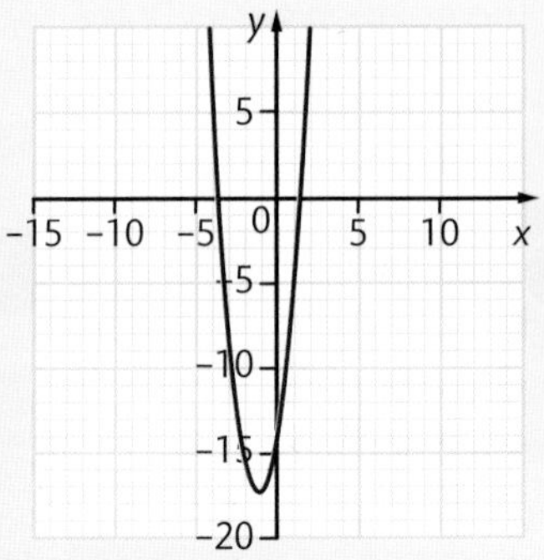

2 The discriminant

In the quadratic formula the expression under the square root sign is called the **discriminant**. The discriminant tells you how many real solutions a quadratic equation has.

$b^2 - 4ac > 0$: equation has two different solutions

$b^2 - 4ac = 0$: equation has one solution

$b^2 - 4ac < 0$: equation has no solutions

2 Worked example Grade 7

Show that the quadratic equation $4x^2 - 3x + 9 = 0$ has no real solutions.

$a = 4 \quad b = -3 \quad c = 9$

Using the discriminant:

$b^2 - 4ac = (-3)^2 - (4 \times 4 \times 9)$

$= 9 - 144$

$= -135 < 0$

There are no real solutions.

15 Exam-style practice Grade 7

1 Solve $\frac{1}{x} - 3x = 4$.

Give your answers correct to 2 decimal places. **[4 marks]**

2 Work out the number of solutions of each quadratic equation.

(a) $6x^2 + 2x - 3 = 0$ **[2 marks]**

(b) $2x^2 - 8x + 11 = 0$ **[2 marks]**

3 The quadratic formula gives the following information

$$x = \frac{13 \pm \sqrt{169 - 80}}{8}$$

Work out the quadratic equation solved.
Give your answer in the form $ax^2 + bx + c = 0$, where a, b and c are integers. **[3 marks]**

Made a start | Feeling confident | Exam ready

Linear inequalities

Inequalities are used to compare values. They show when one value is greater than or less than another value, using the symbols $<$ (is less than), $>$ (is greater than), $\leqslant$ (is less than or equal to) and $\geqslant$ (is greater than or equal to).

⑤ Solving inequalities ✓

Inequalities behave in a similar way to equations. The same rules are used to solve inequalities as for equations. Whatever you do on one side, you must do on the other.

$$2x - 5 < 7 \qquad 2x - 5 + 5 < 7 + 5$$
$$2x < 12 \qquad 2x \div 2 < 12 \div 2$$
$$x < 6$$

The solution set is $\{x \mid x \text{ is a real number and } x < 6\}$.

The aim is to get the unknown on its own, on one side of the inequality, and a number, or an algebraic expression that does not include this unknown, on the other side.

② Worked example — Grade 5 ✓

Given that $-3 < n \leqslant 3$ and n is an integer, write down all the possible values of n.

Values of n are $-2, -1, 0, 1, 2, 3$.
The solution set is $\{n \mid n$ is an integer and $-2, -1, 0, 1, 2\}$.

This is really two inequalities: $n > -3$ and $n \leqslant 3$

Remember that an open circle means $<$ or $>$ and a solid circle means $\leqslant$ or $\geqslant$.

⑤ Worked example — Grade 5 ✓

1 Solve $3(x - 2) > 15$.

$$3x - 6 > 15 \quad \longleftarrow$$ Expand the bracket and then simplify the inequality.
$$3x > 21$$
$$x > 7$$

The solution set is $\{x \mid x$ is a real number and $x > 7\}$.

2

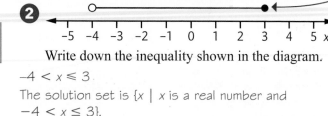

Write down the inequality shown in the diagram.

$$-4 < x \leqslant 3$$

The solution set is $\{x \mid x$ is a real number and $-4 < x \leqslant 3\}$.

② The golden rules ✓

If you multiply or divide by a positive number then the inequality sign does not change.

If you multiply or divide by a negative number then you have to reverse the inequality.

$-x < 10$ multiplied by $-1 = x > -10$

② Inequalities checklist ✓

- ☑ Solve an inequality in exactly the same way as an equation.
- ☑ Multiplying or dividing through by a negative number reverses the sign.
- ☑ $x < 5$ means 'x is less than 5', so 5 is not included in possible values; $x \leqslant 5$ means 5 is included.

⑤ Worked example — Grade 6 ✓

Solve $\dfrac{2 - x}{4} < 3x - 4$.

$$2 - x < 4(3x - 4)$$
$$2 - x < 12x - 16$$
Expand the bracket and then solve the inequality.
$$-x - 12x < -16 - 2$$
$$-13x < -18$$
$$x > \frac{-18}{-13}$$
Reverse the inequality when dividing by a negative number.
$$x > \frac{18}{13}$$

The solution set is $\{x \mid x$ is a real number and $x > \dfrac{18}{13}\}$.

⑩ Exam-style practice — Grades 5–7 ✓

1 Given that x and y are integers such that
$$4 < x < 8 \qquad 5 < y < 10 \qquad x + y = 14$$
list all the possible values of x. **[2 marks]**

2 What are the integer values of x that satisfy both of these inequalities:
$$3x - 12 > 4 \text{ and } 2x - 3 \leqslant 14? \qquad \textbf{[3 marks]}$$

3 (a) Solve the inequality $\dfrac{2x - 3}{3} > \dfrac{5 - x}{2}$. **[3 marks]**

(b) x is an integer. Write down the smallest value of x that satisfies $\dfrac{2x - 3}{3} > \dfrac{5 - x}{2}$. **[1 mark]**

Made a start ✓ | Feeling confident ✓ | Exam ready ✓

Quadratic inequalities

Make sure you are confident solving quadratic equations before you revise quadratic inequalities. For a reminder about quadratic equations, have a look at page 20.

⑤ Solving inequalities

The secret of solving a quadratic inequality is to solve the corresponding quadratic equation and then sketch the graph. For example, to solve:

$x^2 + 5x - 24 \geq 0$

factorise the left-hand side:

$(x + 8)(x - 3) \geq 0$

The **critical values** are $x = -8$ and $x = 3$.

These values occur where the curve with equation $y = (x + 8)(x - 3)$ crosses the x-axis.

Always draw a sketch of your curve, using the critical values to show where the curve cuts the x-axis.

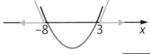

Highlight the area where $y \geq 0$.

$x \leq -8$ and $x \geq 3$ •

The solution set is $\{x \mid x$ is a real number and $-8 \leq x \geq 3\}$.

> There are two regions so write the answer as two inequalities.

⑤ Worked example   Grade 8

Solve the inequality $3(x^2 + 5) < 14x$.

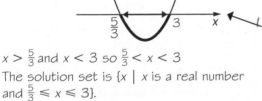

$3x^2 + 15 < 14x$
$3x^2 - 14x + 15 < 0$
$(3x - 5)(x - 3) < 0$
Critical values are $x = \frac{5}{3}$ and $x = 3$.

$x > \frac{5}{3}$ and $x < 3$ so $\frac{5}{3} < x < 3$
The solution set is $\{x \mid x$ is a real number and $\frac{5}{3} \leq x \leq 3\}$.

⑤ Worked example   Grade 8

Solve the inequality $x^2 \leq 16$.

Represent your answer on a number line.

$$x^2 \leq 16$$
$$x^2 - 16 \leq 0$$
$$(x - 4)(x + 4) \leq 0$$

Critical values are $x = -4$ and $x = 4$.

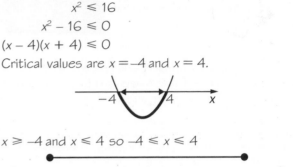

$x \geq -4$ and $x \leq 4$ so $-4 \leq x \leq 4$

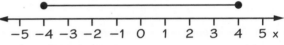

The solution set is $\{x \mid x$ is a real number and $-4 \leq x \leq 4\}$.

② Methods

- ☑ Factorise and solve the quadratic to find the critical values.
- ☑ Sketch a graph.
- ☑ Highlight the area above or below the x-axis.
- ☑ Write out the **two** inequalities.

> First expand the bracket and then rearrange to the form $ax^2 + bx + c < 0$.

> The required region is where the graph is **below** the x-axis.

⑮ Exam-style practice   Grades 5–8

1 (a) Solve the inequality $x^2 - 9 > 0$.   **[2 marks]**

     (b) Represent your answer on a number line.   **[1 mark]**

2 Solve these inequalities.

     (a) $x^2 > 9x - 20$ **[3 marks]**   **(b)** $x^2 > 3(x + 6)$ **[3 marks]**   **(c)** $3x^2 < 2(x + 4)$ **[3 marks]**

3 (a) Sketch the graph of $y = x^2 - 2x - 3$.   **[2 marks]**

     (b) Hence, or otherwise, solve the inequality

       $x^2 - 2x - 3 \geq 0$ **[2 marks]**

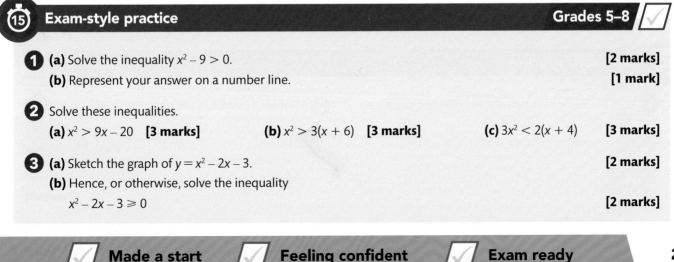

☑ **Made a start**   ☑ **Feeling confident**   ☑ **Exam ready**

Arithmetic sequences

An arithmetic or linear sequence is a sequence of numbers in which the difference between consecutive terms is constant.

5 Finding the *n*th term

An example of an **arithmetic sequence** is

−4 1 6 11 16

+5 +5 +5 +5

The difference between one term and the next term is always +5.

Here is another arithmetic sequence:

8 5 2 −1 −4

−3 −3 −3 −3

You can work out a formula to calculate any term of an arithmetic sequence. This is called the *n*th term, where *n* is an integer, and is of the form $an \pm b$.

A quick method to find *b*

You can use this method to work out the formula for the *n*th term.

*n*th term = difference × *n* + zero term

where zero term = 1st term − difference

Using the zero term is a quick way to find *b*, the number that needs to be added or subtracted to the first part of the formula.

5 Worked example — Grade 5

Here are the first four terms of an arithmetic sequence.

11 17 23 29

Work out, in terms of *n*, an expression for the *n*th term of this arithmetic sequence.

11 17 23 29

+6 +6 +6

Common difference = + 6

Zero term = 11 − 6 = 5

So *n*th term = 6*n* + 5

Check: 3rd term = 23

$6 \times 3 + 5 = 18 + 5 = 23$ ✓

5 Checking a term

You may be asked to work out if a number is part of a given sequence. For example, here are the first five terms of an arithmetic sequence.

3 7 11 15 19

Is 93 a term in the sequence?

Start with the *n*th term.

The *n*th term of this sequence is $4n - 1$.

Set the *n*th term equal to 93 and solve the equation.

$4n - 1 = 93$

$4n = 94$

$n = 23.5$

If your answer is an **integer** (a whole number) then the term is in the sequence. Otherwise, it is not.

93 is not in the sequence.

Always write a conclusion.

5 Worked example — Grade 5

Here are the first five terms of an arithmetic sequence.

5 12 19 26 33

(a) Write down an expression, in terms of *n*, for the *n*th term of the sequence.

5 12 19 26 33

+7 +7 +7 +7

Zero term = 5 − 7 = −2

*n*th term = 7*n* − 2

Work backwards to calculate the zero term of the sequence.

(b) Is 82 a term in the sequence? You must give a reason for your answer.

$7n - 2 = 82$

$7n = 84$

$n = 12$

Yes 82 is in the sequence, because *n* is an integer.

10 Exam-style practice — Grade 5

1 Here are the first five terms of an arithmetic sequence.

8 5 2 −1 −4

Circle the expression for the *n*th term of the sequence.

$3n + 5$ $n - 3$ $11 - 3n$ $8 - 3n$ **[1 mark]**

2 Here are the first five terms of an arithmetic sequence.

3 8 13 18 23

(a) Explain why the number 162 cannot be a term in this sequence. **[1 mark]**

(b) Write down an expression, in terms of *n*, for the *n*th term of the sequence. **[2 marks]**

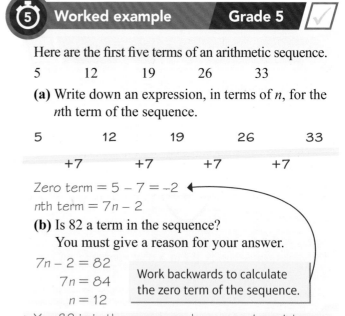

Quadratic sequences

In quadratic sequences the differences between consecutive terms form a linear sequence, meaning that the secondary differences are constant. The formula for the nth term is a quadratic expression, so it includes a term in n^2.

⏱10 Finding the nth term ☑

The nth term of a **quadratic sequence** has the form $an^2 + bn + c$ where a, b and c, are numbers and $a \neq 0$.

The **second differences** of a quadratic sequence are constant.

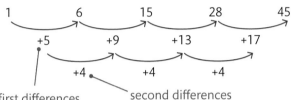

first differences   second differences

Follow the rules to work out the nth term of this sequence.

1 Halve the second difference to get the value of a in $an^2 + bn + c$, so in this case $a = 2$.

second difference $= 4$, $a = 4 \div 2 = 2$

2 Draw a table with 5 rows: n, n^2, an^2, u_n (original sequence term), $u_n - an^2$ (known as the **residue**).

n	1	2	3	4	5
n^2	1	4	9	16	25
$an^2 = 2n^2$	2	8	18	32	50
u_n	1	6	15	28	45
$u_n - an^2$	−1	−2	−3	−4	−5

3 The residue will either be constant or a linear sequence. If it is a linear sequence then find a formula for it in terms of n. Here the residue for the nth term is $-n$.

4 The nth term of the original sequence is $an^2 +$ residue.

nth term $= 2n^2 - n$

⏱5 Worked example   Grade 8 ☑

Here are the first five terms of a quadratic sequence.

1   3   7   13   21

Write down an expression, in terms of n, for the nth term of this quadratic sequence.

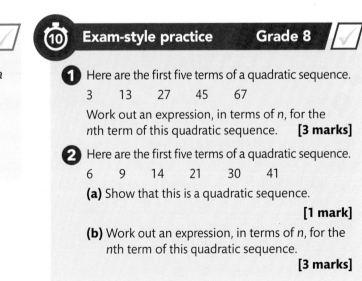

$a = 2 \div 2 = 1$

n	1	2	3	4	5
$an^2 = n^2$	1	4	9	16	25
u_n	1	3	7	13	21
$u_n - an^2$	0	−1	−2	−3	−4

Formula for the residue is: $-n + 1$

nth term $= n^2 - n + 1$

> Always work out the second differences.

> Finding the nth term of this final row, the residue, gives the rest of the quadratic expression for the nth term.

⏱2 Quadratic sequences method ☑

- ☑ Halve the second difference to find the value of a in $an^2 + bn + c$.
- ☑ Draw a table with the five rows as shown above.
- ☑ Work out the formula for the residue.
- ☑ Finally add the formula for the residue to an^2 to get $an^2 + bn + c$.

⏱10 Exam-style practice   Grade 8 ☑

1 Here are the first five terms of a quadratic sequence.

3   13   27   45   67

Work out an expression, in terms of n, for the nth term of this quadratic sequence. **[3 marks]**

2 Here are the first five terms of a quadratic sequence.

6   9   14   21   30   41

(a) Show that this is a quadratic sequence. **[1 mark]**

(b) Work out an expression, in terms of n, for the nth term of this quadratic sequence. **[3 marks]**

Sequence problems

Not all arithmetic sequences are linear or quadratic. You might encounter sequences such as the **Fibonacci sequence**, and sequences involving powers and roots. Arithmetic sequences can often be used to solve mathematical problems.

10 Generating a sequence

To work out the terms of a sequence you can substitute the term numbers into the *n*th term formula.

*n*th term	$2n^2 - 5$	Answer
$n = 1$	$2(1)^2 - 5$	–3
$n = 2$	$2(2)^2 - 5$	3
$n = 3$	$2(3)^2 - 5$	13
$n = 4$	$2(4)^2 - 5$	27

Term-to-term rule

You can use a term-to-term rule to find the next term in a sequence. For example, consider the sequence for which the rule to work out the next term is: 'Add 3 and multiply by 5'.

If the first term is 4 the next term would be

$(4 + 3) \times 5 = 7 \times 5 = 35$

You need to set up algebraic equations.

2 Fibonacci sequences

A Fibonacci sequence is a special sequence that follows a certain rule.

The rule for generating the Fibonacci sequence is: 'Add two consecutive terms to get the next term.'

Fibonacci sequences can have different starting values. Here are two examples:

2 4 6 10 16 26

1 5 6 11 17 28

You must show your working clearly.

5 Worked example — Grades 5–6

1 The *n*th term of a sequence is $(\sqrt{5})^n$.

Work out the first three terms of the sequence.

$n = 1$	$(\sqrt{5})^1$	$\sqrt{5}$
$n = 2$	$(\sqrt{5})^2 = \sqrt{5} \times \sqrt{5}$	5
$n = 3$	$(\sqrt{5})^3 = \sqrt{5} \times \sqrt{5} \times \sqrt{5}$	$5\sqrt{5}$

2 The rule for finding the next term in a sequence is: 'Subtract *k* then multiply by 4'.

The second term is 16 and the third term is 28.

Work out:

(a) the value of *k*

Work backwards using the term-to-term rule.

$28 = 4(16 - k)$

$28 = 64 - 4k$

$4k = 36$, so $k = 9$

(b) the first term.

Substitute the value of *k*.

$16 = 4(a - 9)$

$16 = 4a - 36$

$4a = 16 + 36 = 52$, so $a = 13$

2 Worked example — Grade 6

Work out the next three terms of this Fibonacci sequence.

1 1 2 3

fifth term: $2 + 3 = 5$

sixth term: $3 + 5 = 8$

seventh term: $5 + 8 = 13$

10 Exam-style practice — Grade 5

1 The *n*th term of a sequence is $an + b$ where *a* and *b* are integers. The third term is 14 and the seventh term is 26. What are the values of *a* and *b*? **[3 marks]**

2 The first three terms of a Fibonacci-type sequence are

$a \quad b \quad a + b$

(a) Show that the sixth term of this sequence is $3a + 5b$. **[2 marks]**

(b) Given that the third term is 4 and the sixth term is 18, work out the value of *a* and the value of *b*. **[3 marks]**

3 The *n*th term of an arithmetic sequence is given by $u_n = 4n - 1$, where *n* is an integer.

(a) Determine whether 53 is a term in this sequence. **[1 mark]**

(b) Write an expression for u_n in terms of u_{n-1}. **[2 marks]**

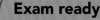

Drawing straight-line graphs

You need to know how to draw a straight-line graph and how to find the equation of a given straight line.

(5) Straight-line graphs

A straight line graph has the equation $y = mx + c$ where m is the **gradient** or slope of the line and c is the **y-intercept**, where the line cuts the y-axis.

This graph has the equation $y = 4x - 6$. This means that the gradient is 4 and the y-coordinate of the y-intercept is –6, so the y-intercept is at (0, –6).

You can find the equation of a graph by working out the gradient and the y-intercept.

> To find the gradient and the y-intercept always rearrange the equation in the form $y = mx + c$.

> Use the grid lines to draw a right-angled triangle. Remember that the gradient is the vertical length divided by the horizontal length.

(5) Worked example — Grades 4–5

1 The equation of a straight line is $3y + 8x = 7$.

(a) Write down the gradient.

$3y + 8x = 7$
$3y = -8x + 7$
$y = -\frac{8}{3}x + \frac{7}{3}$

Gradient $= -\frac{8}{3}$

(b) Write down the coordinates of the y-intercept.

Coordinates are $\left(\frac{7}{3}, 0\right)$

2 Work out the equation of the straight line shown.

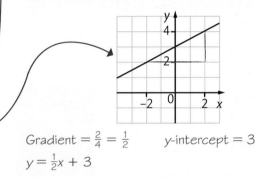

Gradient $= \frac{2}{4} = \frac{1}{2}$ y-intercept $= 3$

$y = \frac{1}{2}x + 3$

(2) Drawing straight-line graphs

You can use a table of values to draw a straight-line graph.

1 Choose three simple points as the x-values.

2 Substitute these values into the equation to find the y-values, and add to your table of values.

3 Plot the points on the graph and join them with a straight line.

(1) Straight-line graphs checklist

☑ Draw a triangle to find the gradient, m.

☑ Look at the y-coordinate when the graph crosses the y-axis to find c.

☑ Put your values of m and c into the equation $y = mx + c$.

(2) Worked example — Grade 4

Draw a graph of $x + y = 5$ for values of x from –2 to 4.

x	–2	0	4
$y = 5 - x$	7	5	1

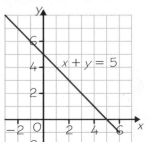

(10) Exam-style practice — Grades 4–5

1 Draw the graph of $y = 3 - 2x$ for values of x from –2 to 3. **[3 marks]**

2 The diagram shows three points A (–1, 5), B (2, –1) and C (0, 5).

Line L is parallel to AB and passes through C.

Work out the equation of the line L.

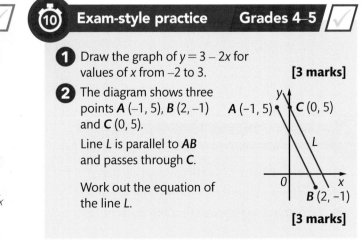

[3 marks]

Equations of straight lines

You should be able to use algebra to work out the equation of a straight line.

⑤ One point and a gradient ☑

You need to be able to use the general form $y = mx + c$, where m is the gradient and c is the y-intercept, to find the equation of a straight line.

For example, given that a line passes through the point $(3, 7)$ and has gradient 4, you can substitute the values of x, y and m into the equation to work out the value of c.

$y = mx + c$

$7 = 4(3) + c$

$7 = 12 + c$

Rearranging, $c = 7 - 12 = -5$.

Hence, $y = 4x - 5$.

⑤ Finding the gradient ☑

If you are given two points, you can work out the gradient as: $\dfrac{\text{change in } y}{\text{change in } x}$ then proceed as before.

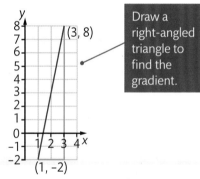

Draw a right-angled triangle to find the gradient.

For example, for the straight line passing through $(1, -2)$ and $(3, 8)$:

$$\dfrac{\text{change in } y}{\text{change in } x} = \dfrac{8 - (-2)}{3 - 1} = \dfrac{10}{2} = 5$$

Now substitute the x- and y-values for one of the points on the line to work out the equation of the line:

$y = mx + c$

$-2 = 5(1) + c$ $8 = 5(3) + c$

$-2 = 5 + c$ $8 = 15 + c$

$c = -7$ $c = -7$

$y = 5x - 7$ $y = 5x - 7$

You get the same value of c using either point on the line.

① Straight-line equation checklist ☑

- ☑ The equation of a straight line is $y = mx + c$.
- ☑ To work out c substitute correctly for x, y and m.
- ☑ If 2 points are given work out the gradient first.

⑤ Worked example   Grade 6 ☑

Work out the equation of the line that passes through $(4, -7)$ and has gradient $-\frac{1}{2}$.

$y = mx + c$

$-7 = (-\frac{1}{2})(4) + c$ ← Substitute the values into the equation.

$-7 = -2 + c$

$c = -5$

Hence, $y = -\frac{1}{2}x - 5$

⑤ Worked example   Grade 6 ☑

Work out the equation of the line that passes through the points $(-9, 4)$ and $(-6, 2)$.

$$\text{Gradient} = \dfrac{\text{change in } y}{\text{change in } x}$$

$$= \dfrac{4 - 2}{-9 - (-6)}$$

$$= \dfrac{2}{-9 + 6} = -\dfrac{2}{3}$$

Remember that the gradient is equal to the vertical rise divided by the horizontal distance.

$y = mx + c$

$2 = (-\frac{2}{3})(-6) + c$

$2 = 4 + c$

$c = -2$

Hence, $y = -\frac{2}{3}x - 2$

Use brackets when you substitute negative numbers.

⑩ Exam-style practice   Grade 6 ☑

① Work out the equation of the line with gradient -3 that passes through the point $(2, 5)$.

[2 marks]

② The line L passes through the points $(0, -2)$ and $(6, 1)$. Work out the equation of the line L.

[3 marks]

③ The point A has coordinates $(0, 2)$. The point B has coordinates $(-4, -1)$.

Work out the equation of the line AB.   **[3 marks]**

☑ Made a start   ☑ Feeling confident   ☑ Exam ready

Parallel and perpendicular lines

If two lines are perpendicular, they always meet or cross at a right angle. If two lines are parallel, they never meet, but the perpendicular distance between them is constant.

2 Parallel lines

Parallel lines, such as L_1 and L_2, have the same gradient.

For example, the lines $y = 5x + 6$ and $y = 5x - 7$ are parallel, as the gradient is 5 in both of them.

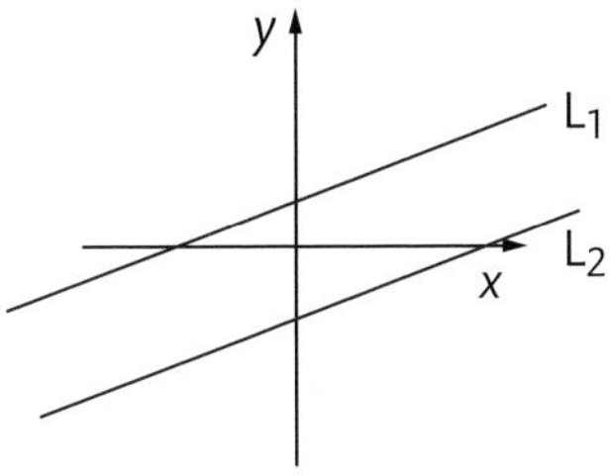

2 Perpendicular lines

Two lines are **perpendicular** when they meet at a right angle or 90°.

The product of the gradients of two perpendicular lines is –1.

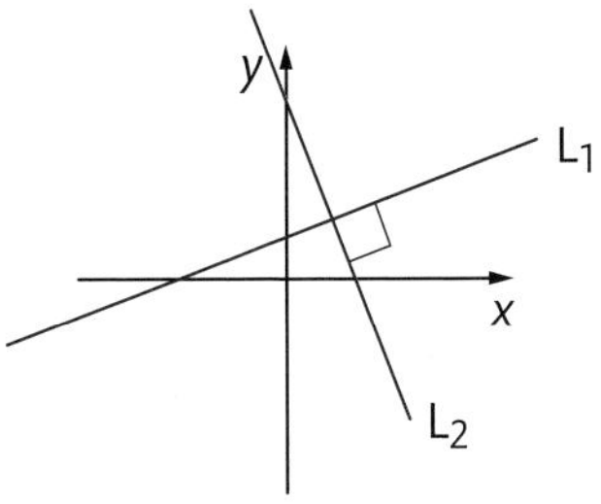

10 Worked example — Grade 7

1 Work out the equation of the line that is perpendicular to the line $y = 5x + 6$ and passes through the point (5, –2).

Gradient of $y = 5x + 6$ is 5.

Gradient of line perpendicular to $y = 5x + 6$ is $-\frac{1}{5}$.

Line through (5, –2), with gradient $-\frac{1}{5}$:

$-2 = -\frac{1}{5}(5) + c$

$-2 = -1 + c$

$c = -1$ Hence, $y = -\frac{1}{5}x - 1$.

2 A is the point with coordinates (3, –1).

B is the point with coordinates (6, 5).

The straight line L goes through both A and B.

Is the line with equation $3y = 2x - 4$ perpendicular to line L? You must show your working.

Gradient of AB $= \frac{5 - (-1)}{6 - 3} = \frac{6}{3} = 2$

Rearrange equation $3y = 2x - 4$: $y = \frac{2}{3}x - \frac{4}{3}$

Gradient of line $= \frac{2}{3}$

$m_1 \times m_2 = 2 \times \frac{2}{3} = \frac{4}{3} \neq -1$

The lines are not perpendicular because the product of the gradients is not equal to –1.

5 Worked example — Grade 7

L_1 and L_2 are parallel lines. The equation of L_1 is $y = 3x + 1$. L_2 passes through the point (5, 4).

Write down an equation for L_2.

The gradient of L_1 is 3 so the gradient of L_2 is 3.

Using $y = mx + c$ for the line through (5, 4), with gradient 3, L_2 has y-intercept:

$4 = 3(5) + c$

$4 = 15 + c$

$c = -11$

Hence, $y = 3x - 11$.

To work out the equation of a straight line look back at pages 29 and 30.

The gradient of the perpendicular is the **negative reciprocal** of the original gradient, which means that $m_1 \times m_2 = -1$.

Problem solving

For lines to be perpendicular the product of the two gradients must be –1. You can use this fact to check whether two lines are perpendicular.

1 Checklist

- Parallel lines have the same gradient, $m_1 = m_2$.
- For two lines that are perpendicular, $m_1 \times m_2 = -1$.
- To work out the equations of parallel or perpendicular lines, substitute values into $y = mx + c$.

10 Exam-style practice — Grade 7

1 The straight line *L* has the equation $y = 3x - 4$.

(a) Write down the equation of the line parallel to *L* which passes through the origin. **[2 marks]**

(b) Work out the equation of the straight line that passes through (0, 5) and is perpendicular to *L*. **[2 marks]**

2 The straight line *L* has the equation $y = 2x - 5$.

Work out the equation of the straight line perpendicular to *L* which passes through (–2, 3). **[3 marks]**

3 *A* and *B* are straight lines. Line *A* has the equation $2y = 3x + 8$. Line *B* goes through the points (–1, 2) and (2, 8). Determine whether lines *A* and *B* intersect. You must show your working. **[3 marks]**

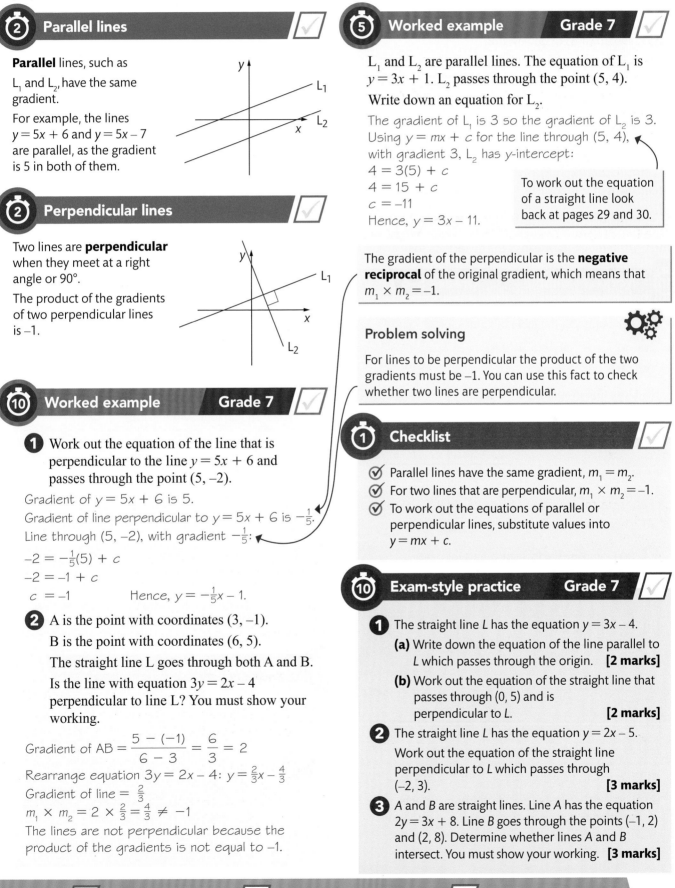

Quadratic graphs

The equation for a quadratic graph is a quadratic equation, with an x^2 term in it. You should be able to draw a quadratic graph from a set of values and recognise the line of symmetry.

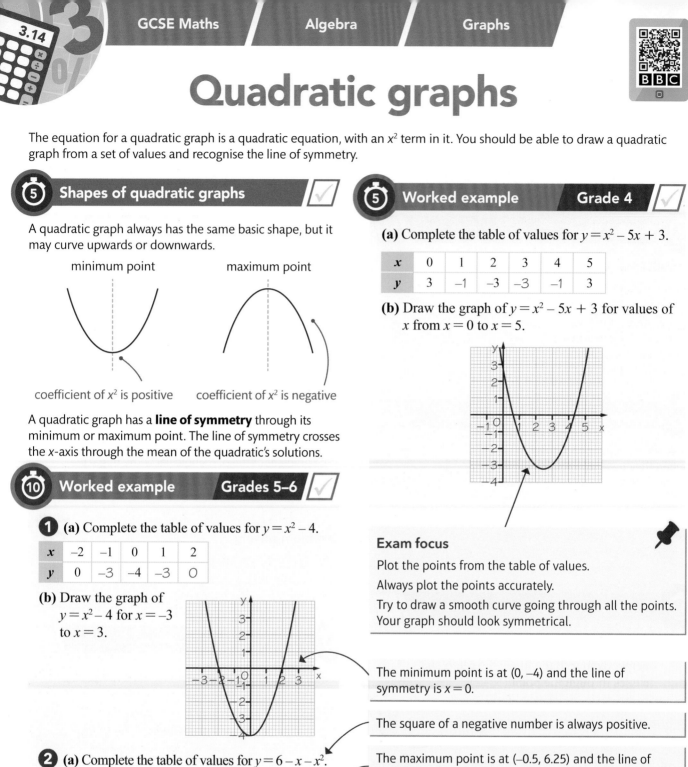

5 Shapes of quadratic graphs

A quadratic graph always has the same basic shape, but it may curve upwards or downwards.

minimum point | maximum point

coefficient of x^2 is positive | coefficient of x^2 is negative

A quadratic graph has a **line of symmetry** through its minimum or maximum point. The line of symmetry crosses the x-axis through the mean of the quadratic's solutions.

10 Worked example — Grades 5–6

1 **(a)** Complete the table of values for $y = x^2 - 4$.

x	−2	−1	0	1	2
y	0	−3	−4	−3	0

(b) Draw the graph of $y = x^2 - 4$ for $x = -3$ to $x = 3$.

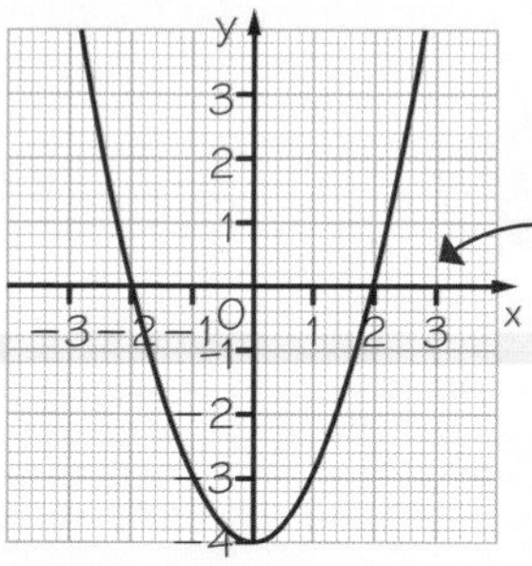

The minimum point is at (0, −4) and the line of symmetry is $x = 0$.

The square of a negative number is always positive.

2 **(a)** Complete the table of values for $y = 6 - x - x^2$.

x	−3	−2	−1	0	1	2
y	0	4	6	6	4	0

(b) Draw the graph of $y = 6 - x - x^2$ for $x = -4$ to $x = 3$.

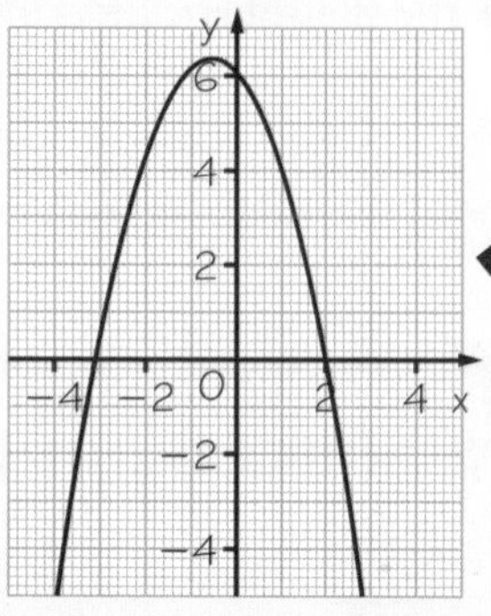

The maximum point is at (−0.5, 6.25) and the line of symmetry is $x = -0.5$.

(c) Work out estimates for the solutions of the equation $6 - x - x^2 = 0$.

$x = -3$ and $x = 2$

The solutions to the equation are the x-coordinates at the point where the curve crosses the x-axis.

5 Worked example — Grade 4

(a) Complete the table of values for $y = x^2 - 5x + 3$.

x	0	1	2	3	4	5
y	3	−1	−3	−3	−1	3

(b) Draw the graph of $y = x^2 - 5x + 3$ for values of x from $x = 0$ to $x = 5$.

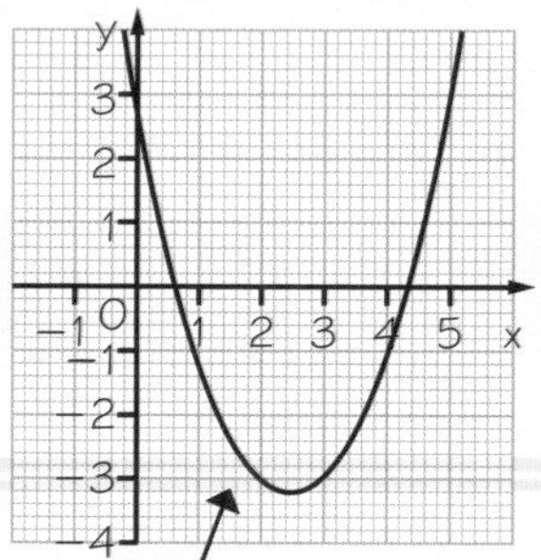

Exam focus

Plot the points from the table of values.

Always plot the points accurately.

Try to draw a smooth curve going through all the points. Your graph should look symmetrical.

10 Exam-style practice — Grades 5–6

(a) Draw a table of values for $y = x^2 - 3x + 2$ from $x = -1$ to $x = 4$. **[2 marks]**

(b) On a grid, draw the graph of $y = x^2 - 3x + 2$ for values of x from −1 to 4. **[2 marks]**

(c) Work out estimates for the solutions of the equation $x^2 - 3x + 2 = 0$. **[2 marks]**

Cubic and reciprocal graphs

You should be able to recognise, draw and interpret the shapes of cubic and reciprocal graphs.

② Cubic graphs

In a cubic expression, the highest power of x is x^3.
Here are two examples of cubic graphs.

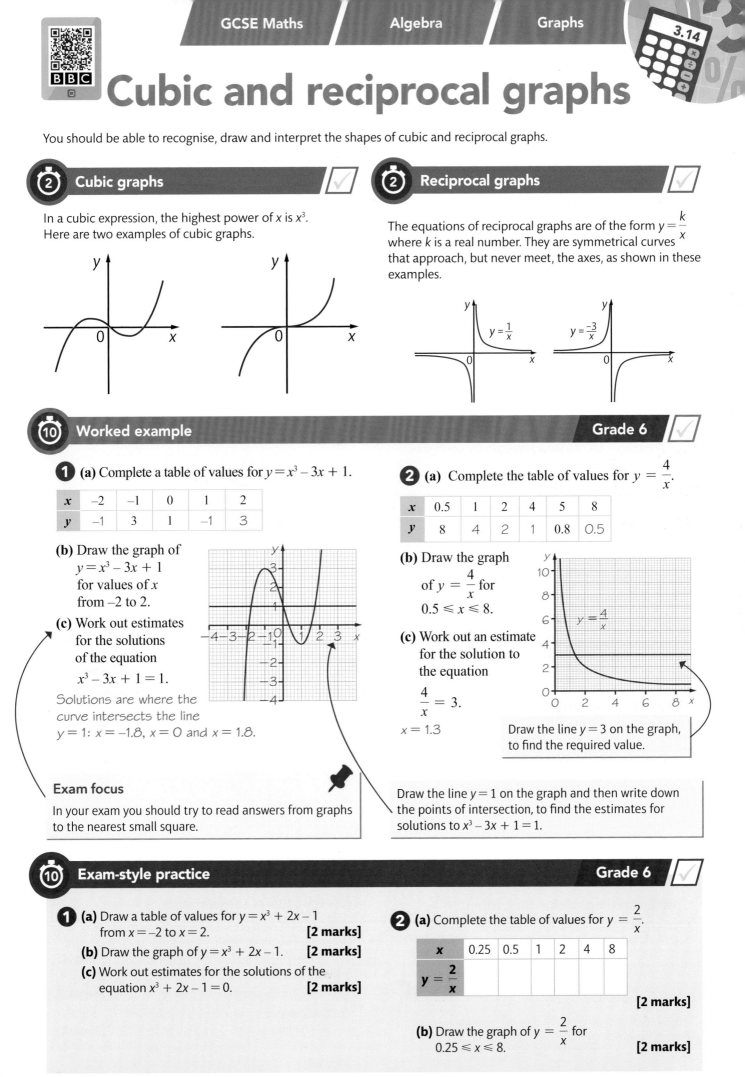

② Reciprocal graphs

The equations of reciprocal graphs are of the form $y = \dfrac{k}{x}$ where k is a real number. They are symmetrical curves that approach, but never meet, the axes, as shown in these examples.

$y = \dfrac{1}{x}$ $y = \dfrac{-3}{x}$

⑩ Worked example   Grade 6

① **(a)** Complete a table of values for $y = x^3 - 3x + 1$.

x	−2	−1	0	1	2
y	−1	3	1	−1	3

(b) Draw the graph of $y = x^3 - 3x + 1$ for values of x from −2 to 2.

(c) Work out estimates for the solutions of the equation $x^3 - 3x + 1 = 1$.

Solutions are where the curve intersects the line $y = 1$: $x = -1.8$, $x = 0$ and $x = 1.8$.

② **(a)** Complete the table of values for $y = \dfrac{4}{x}$.

x	0.5	1	2	4	5	8
y	8	4	2	1	0.8	0.5

(b) Draw the graph of $y = \dfrac{4}{x}$ for $0.5 \leqslant x \leqslant 8$.

(c) Work out an estimate for the solution to the equation $\dfrac{4}{x} = 3$.

$x = 1.3$

Draw the line $y = 3$ on the graph, to find the required value.

Exam focus

In your exam you should try to read answers from graphs to the nearest small square.

Draw the line $y = 1$ on the graph and then write down the points of intersection, to find the estimates for solutions to $x^3 - 3x + 1 = 1$.

⑩ Exam-style practice   Grade 6

① **(a)** Draw a table of values for $y = x^3 + 2x - 1$ from $x = -2$ to $x = 2$. **[2 marks]**

(b) Draw the graph of $y = x^3 + 2x - 1$. **[2 marks]**

(c) Work out estimates for the solutions of the equation $x^3 + 2x - 1 = 0$. **[2 marks]**

② **(a)** Complete the table of values for $y = \dfrac{2}{x}$.

x	0.25	0.5	1	2	4	8
$y = \dfrac{2}{x}$						

[2 marks]

(b) Draw the graph of $y = \dfrac{2}{x}$ for $0.25 \leqslant x \leqslant 8$. **[2 marks]**

✓ **Made a start**   ✓ **Feeling confident**   ✓ **Exam ready**

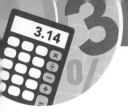

Real-life graphs

Graphs can model real-life situations and display information. You need to know about conversion graphs, distance-time graphs and speed-time graphs.

5 Rate of change

If a graph has **time** on its horizontal axis, then the gradient of the graph represents a **rate of change**. For example on a distance–time graph the gradient represents **speed**, and on a speed–time graph the gradient represents **acceleration**.

Draw a triangle, and use the scale when working out the lengths of its sides.

The gradient is negative because the slope of the line is downwards.

Exam focus

When a question is based on a real-life example, you should explain what any values mean in the context of the question.

10 Worked example — Grade 5

Look at this distance–time graph.

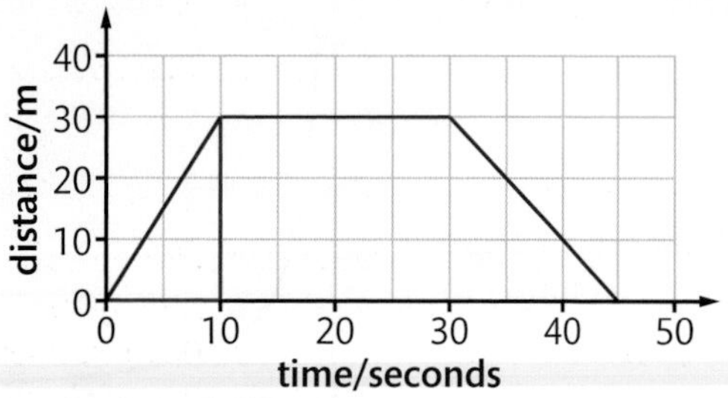

(a) (i) Work out the gradient of this distance–time graph in the first 10 seconds.

$$\text{Gradient} = \frac{30 - 0}{10 - 0} = \frac{30}{10} = 3$$

(ii) Interpret the value of the gradient.

The gradient represents speed, so the speed is 3 m/s.

(b) (i) Write down the gradient between 10 and 30 seconds.

Gradient = 0

(ii) Interpret the value of the gradient.

A zero gradient means that the object is not moving – it is stationary.

Exam focus

'Interpret' means to describe how this value relates to the motion of the object.

5 Worked example — Grade 5

Petrol is being pumped out of a tank. The graph shows the depth, d cm, of petrol in the tank after t minutes.

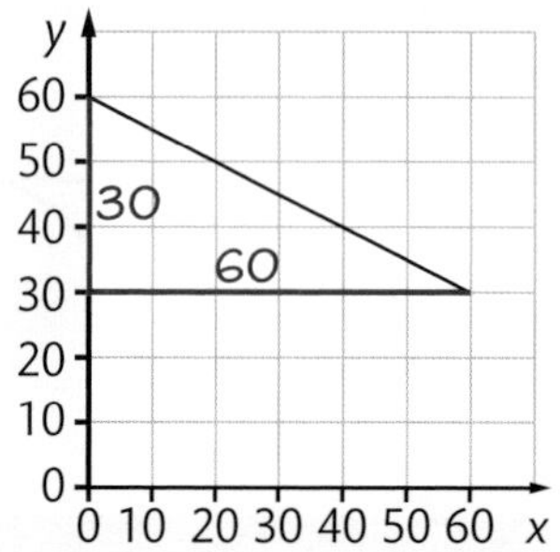

(a) Work out the gradient of this graph.

$$\text{Gradient} = \frac{\text{change in } y}{\text{change in } x} = \frac{-30}{60} = -0.5$$

(b) Explain what this gradient represents.

The gradient represents the rate of change of the depth of the petrol in the tank.

5 Exam-style practice — Grade 5

Water is leaking out of two containers, **A** and **B**. The water started to leak out of the containers at the same time.

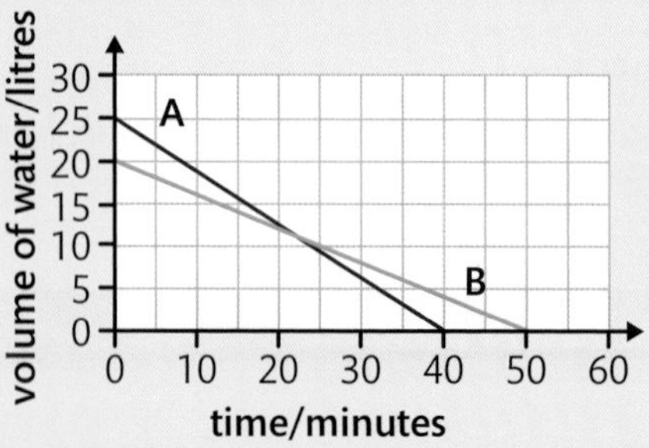

The straight line **A** shows information about the amount of water, in litres, in container **A**. The straight line **B** shows information about the amount of water, in litres, in container **B**.

(a) Work out the gradient of line **A**. **[2 marks]**

(b) State, with a reason, which container will be empty first. **[1 mark]**

Trigonometric graphs

Sin x, cos x and tan x are all trigonometrical functions and they each have their own distinctive graphs. You need to be familiar with the shapes of these graphs.

5 The three graphs

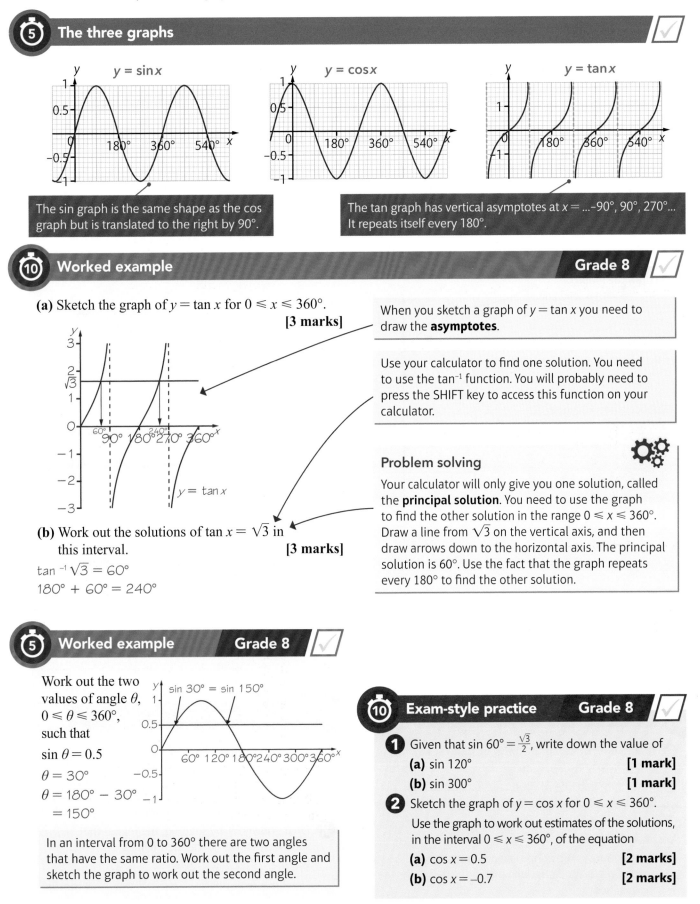

The sin graph is the same shape as the cos graph but is translated to the right by 90°.

The tan graph has vertical asymptotes at $x = \dots$–90°, 90°, 270°... It repeats itself every 180°.

10 Worked example

Grade 8

(a) Sketch the graph of $y = \tan x$ for $0 \leqslant x \leqslant 360°$.

[3 marks]

$y = \tan x$

When you sketch a graph of $y = \tan x$ you need to draw the **asymptotes**.

Use your calculator to find one solution. You need to use the $\tan^{-1}$ function. You will probably need to press the SHIFT key to access this function on your calculator.

Problem solving

Your calculator will only give you one solution, called the **principal solution**. You need to use the graph to find the other solution in the range $0 \leqslant x \leqslant 360°$. Draw a line from $\sqrt{3}$ on the vertical axis, and then draw arrows down to the horizontal axis. The principal solution is 60°. Use the fact that the graph repeats every 180° to find the other solution.

(b) Work out the solutions of $\tan x = \sqrt{3}$ in this interval. **[3 marks]**

$\tan^{-1} \sqrt{3} = 60°$

$180° + 60° = 240°$

5 Worked example

Grade 8

Work out the two values of angle θ, $0 \leqslant \theta \leqslant 360°$, such that

$\sin \theta = 0.5$

$\theta = 30°$

$\theta = 180° - 30°$

$= 150°$

$\sin 30° = \sin 150°$

In an interval from 0 to 360° there are two angles that have the same ratio. Work out the first angle and sketch the graph to work out the second angle.

10 Exam-style practice

Grade 8

1 Given that $\sin 60° = \frac{\sqrt{3}}{2}$, write down the value of

(a) $\sin 120°$ **[1 mark]**

(b) $\sin 300°$ **[1 mark]**

2 Sketch the graph of $y = \cos x$ for $0 \leqslant x \leqslant 360°$.

Use the graph to work out estimates of the solutions, in the interval $0 \leqslant x \leqslant 360°$, of the equation

(a) $\cos x = 0.5$ **[2 marks]**

(b) $\cos x = -0.7$ **[2 marks]**

Inequalities on graphs

Inequalities can be used to represent **regions** on graphs. You need to be able to shade a region that satisfies a given set of inequalities.

② Solid or dotted lines

If the inequality is $<$ or $>$ draw a dotted line - - -.

If the inequality is $\leqslant$ or $\geqslant$ draw a solid line ——.

> Dotted lines mean that the points on the line are not included, solid lines mean that the points on the line are included.

⑩ Worked example   Grade 7

On the grid below show, by shading, the region defined by the inequalities

$$x + y \leqslant 5 \qquad y \geqslant x - 2 \qquad x > 1$$

Mark this region with the letter **R**.

Exam focus

Make sure you can draw straight-line graphs. See page 29 to recap drawing up tables of values and plotting the points, if you need to.

The inequality sign is $\leqslant$ so draw the graph of $x + y = 5$ with a solid line.

The inequality sign is $\geqslant$ so draw the graph of $y = x - 2$ with a solid line.

Choose three x-values and work out the corresponding y-values.
$x = 0, y = 5$
$x = 2, y = 3$
$x = 5, y = 0$

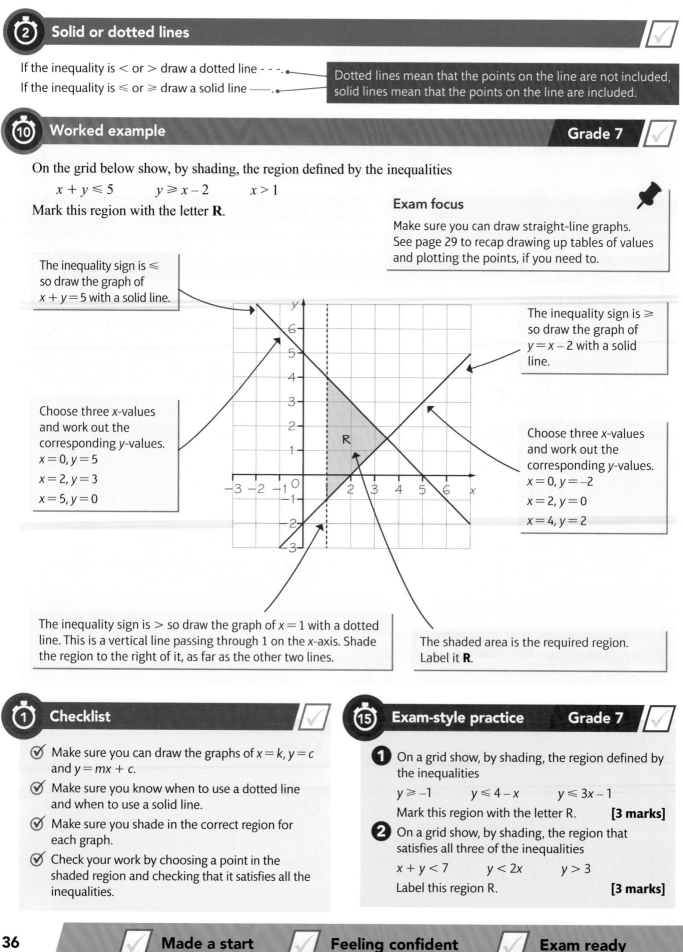

Choose three x-values and work out the corresponding y-values.
$x = 0, y = -2$
$x = 2, y = 0$
$x = 4, y = 2$

The inequality sign is $>$ so draw the graph of $x = 1$ with a dotted line. This is a vertical line passing through 1 on the x-axis. Shade the region to the right of it, as far as the other two lines.

The shaded area is the required region. Label it **R**.

① Checklist

☑ Make sure you can draw the graphs of $x = k$, $y = c$ and $y = mx + c$.

☑ Make sure you know when to use a dotted line and when to use a solid line.

☑ Make sure you shade in the correct region for each graph.

☑ Check your work by choosing a point in the shaded region and checking that it satisfies all the inequalities.

⑮ Exam-style practice   Grade 7

1 On a grid show, by shading, the region defined by the inequalities

$$y \geqslant -1 \qquad y \leqslant 4 - x \qquad y \leqslant 3x - 1$$

Mark this region with the letter R.   **[3 marks]**

2 On a grid show, by shading, the region that satisfies all three of the inequalities

$$x + y < 7 \qquad y < 2x \qquad y > 3$$

Label this region R.   **[3 marks]**

☑ **Made a start**   ☑ **Feeling confident**   ☑ **Exam ready**

Using quadratic graphs

When you have to solve simultaneous equations, one linear and one quadratic, you can use quadratic graphs.

10 Solving simultaneous equations

To solve simultaneous equations that involve a linear and a quadratic equation, draw the quadratic graph and the straight line on the same grid. The solutions are where the straight line cuts the curve. The points were the curve crosses the *x*-axis are called roots.

For example, the diagram shows the graph of $y = x^2 - 6x - 2$.

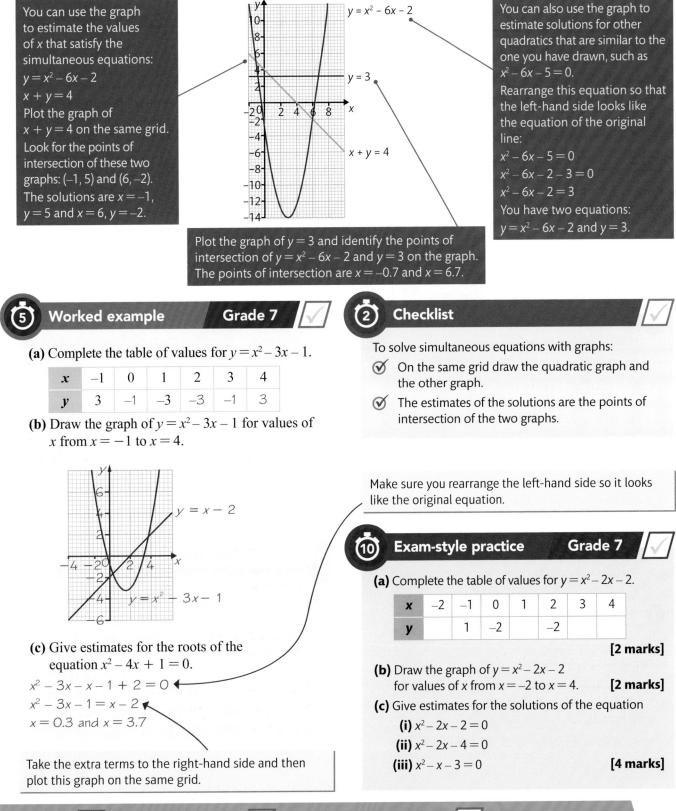

You can use the graph to estimate the values of *x* that satisfy the simultaneous equations:
$y = x^2 - 6x - 2$
$x + y = 4$
Plot the graph of $x + y = 4$ on the same grid.
Look for the points of intersection of these two graphs: (–1, 5) and (6, –2).
The solutions are $x = -1$, $y = 5$ and $x = 6$, $y = -2$.

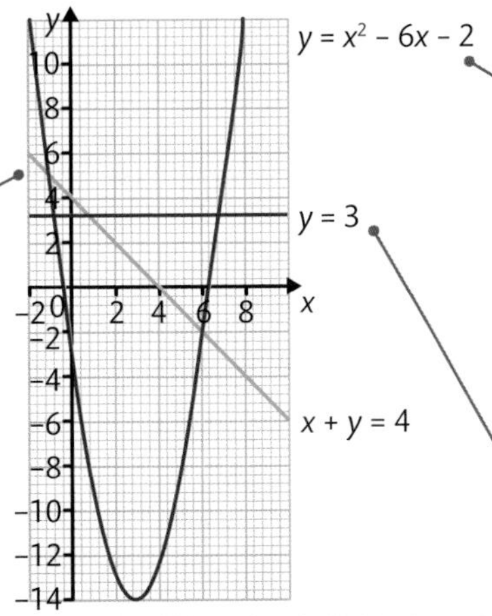

Plot the graph of $y = 3$ and identify the points of intersection of $y = x^2 - 6x - 2$ and $y = 3$ on the graph. The points of intersection are $x = -0.7$ and $x = 6.7$.

You can also use the graph to estimate solutions for other quadratics that are similar to the one you have drawn, such as $x^2 - 6x - 5 = 0$.
Rearrange this equation so that the left-hand side looks like the equation of the original line:
$x^2 - 6x - 5 = 0$
$x^2 - 6x - 2 - 3 = 0$
$x^2 - 6x - 2 = 3$
You have two equations:
$y = x^2 - 6x - 2$ and $y = 3$.

5 Worked example — Grade 7

(a) Complete the table of values for $y = x^2 - 3x - 1$.

x	–1	0	1	2	3	4
y	3	–1	–3	–3	–1	3

(b) Draw the graph of $y = x^2 - 3x - 1$ for values of *x* from $x = -1$ to $x = 4$.

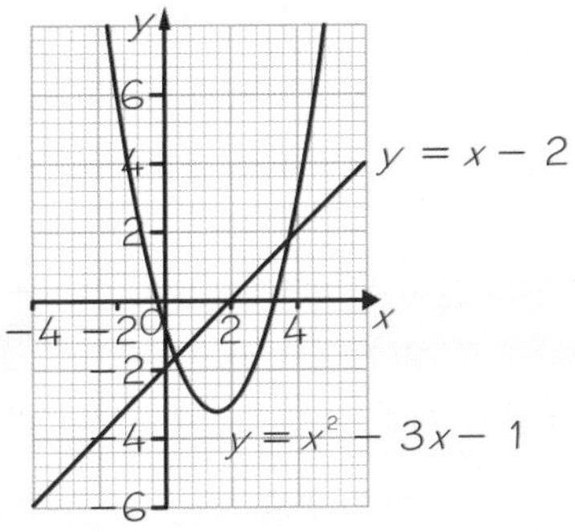

(c) Give estimates for the roots of the equation $x^2 - 4x + 1 = 0$.

$x^2 - 3x - x - 1 + 2 = 0$
$x^2 - 3x - 1 = x - 2$
$x = 0.3$ and $x = 3.7$

Make sure you rearrange the left-hand side so it looks like the original equation.

Take the extra terms to the right-hand side and then plot this graph on the same grid.

2 Checklist

To solve simultaneous equations with graphs:

- ✓ On the same grid draw the quadratic graph and the other graph.
- ✓ The estimates of the solutions are the points of intersection of the two graphs.

10 Exam-style practice — Grade 7

(a) Complete the table of values for $y = x^2 - 2x - 2$.

x	–2	–1	0	1	2	3	4
y		1	–2		–2		

[2 marks]

(b) Draw the graph of $y = x^2 - 2x - 2$ for values of *x* from $x = -2$ to $x = 4$. **[2 marks]**

(c) Give estimates for the solutions of the equation
(i) $x^2 - 2x - 2 = 0$
(ii) $x^2 - 2x - 4 = 0$
(iii) $x^2 - x - 3 = 0$ **[4 marks]**

Turning points

You need to be able to find the turning point of a quadratic graph. This is the point at which the graph changes direction.

② Shapes of turning points

Quadratic graphs have a **minimum** turning point if the coefficient of x^2 is positive, and a **maximum** turning point if the coefficient of x^2 is negative.

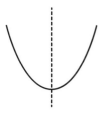 

minimum turning point   maximum turning point

You need to use the method of completing the square (see page 22) to determine the turning points of a quadratic curve.

When an equation is written as $y = (x - b)^2 + c$ then this is the graph of $y = x^2$ translated b units right along the x-axis and c units up, from which you can say that the coordinates of the turning point are (b, c).

The minimum or maximum value of y is c.

⑩ Worked example — Grade 9

1 (a) Write $x^2 - 6x - 5$ in the form $(x + a)^2 + b$ where a and b are integers.

$$x^2 - 6x - 5 = (x - 3)^2 - 3^2 - 5$$
$$= (x - 3)^2 - 9 - 5$$
$$= (x - 3)^2 - 14$$

(b) Hence, or otherwise, write down the coordinates of the turning point of the graph of $y = x^2 - 6x - 5$.

$(3, -14)$

2 Given that the minimum turning point of a quadratic curve is $(3, -5)$, find the equation of the curve in the form $y = x^2 + px + q$.

$$y = (x - 3)^2 - 5$$
$$= x^2 - 3x - 3x + 9 - 5$$
$$= x^2 - 6x + 4$$

3 (a) Write $2x^2 + 16x + 25$ in the form $a(x + b)^2 + c$ where a, b, and c are integers.

$$2x^2 + 16x + 25 = 2[x^2 + 8x] + 25$$
$$= 2[(x + 4)^2 - 16] + 25$$
$$= 2(x + 4)^2 - 32 + 25$$
$$= 2(x + 4)^2 - 7$$

(b) Hence, or otherwise, write down the minimum value of $2x^2 + 16x + 25$.

Minimum value is -7.

② Worked example — Grade 9

The graph of $y = f(x)$ is drawn on the grid.

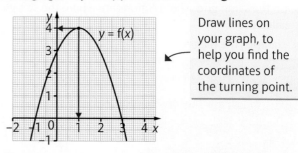

Draw lines on your graph, to help you find the coordinates of the turning point.

Write down the coordinates of the turning point of the graph.

$(1, 4)$

① Checklist

☑ Make sure you can complete the square.

☑ If $y = a(x + b)^2 + c$ then the turning point is $(-b, c)$.

Complete the square.

You may need to revise the method of completing the square. Look on page 22.

Use the rearranged equation to write down the turning point.

Problem solving

$2(x + 4)^2$ must always be positive, so the minimum value it can take is 0. This means that the whole expression has a minimum value of -7, which occurs when $x = -4$.

⑮ Exam-style practice — Grade 9

1 (a) Write $2x^2 + 16x + 33$ in the form $a(x + b)^2 + c$ where a, b, and c are integers. **[3 marks]**

(b) Hence, or otherwise, write down the coordinates of the turning point of the graph of $y = 2x^2 + 16x + 33$. **[1 mark]**

2 Write down the turning points of the following curves.

(a) $y = x^2 - 10x + 17$ **[3 marks]**

(b) $y = 4x^2 - 24x - 15$ **[4 marks]**

(c) $y = 9 - 8x - x^2$ **[4 marks]**

Made a start | Feeling confident | Exam ready

3.14

Sketching graphs

You can sketch the graphs of quadratic and cubic functions if you know a few basic steps.

② Sketching quadratics

Follow these steps to sketch a quadratic graph.

1 Write out your quadratic in the form $a(x + b)^2 + c$.

2 Write down the turning point.

3 Determine the shape of the graph by looking at the coefficient of x^2.

4 Work out the roots, if possible, by letting $a(x + b)^2 + c = 0$, to find where the graph cuts the x-axis.

> Complete the square to determine the turning point.

> The roots are any intercepts with the x-axis. Work them out by solving the equation.

② Sketching cubics

You can sketch cubic graphs by following these steps.

1 Write the cubic equation in the form $(x + a)(x + b)(x + c) = 0$ to obtain the roots of the equation.

2 Look at the coefficient of x^3 to determine the shape of the graph.

① Sketching graphs checklist

☑ When sketching quadratic graphs write out the equation in the form $y = a(x + b)^2 + c$ to determine the turning points.

☑ When sketching cubic graphs write out the equation in the form $y = (x + a)(x + b)(x + c)$.

⑩ Worked example — Grades 8–9

1 Sketch the graph of $y = x^2 - 10x + 21$, showing the coordinates of the turning point and the coordinates of any roots.

$$x^2 - 10x + 21 = (x - 5)^2 - 5^2 + 21$$
$$= (x - 5)^2 - 25 + 21$$
$$= (x - 5)^2 - 4$$

Turning point is $(5, -4)$.
To work out the roots:
$$(x - 5)^2 - 4 = 0$$
$$(x - 5)^2 = 4$$
$$x - 5 = \pm 2$$
$$x = 5 \pm 2$$
$$x = 3 \text{ or } x = 7$$

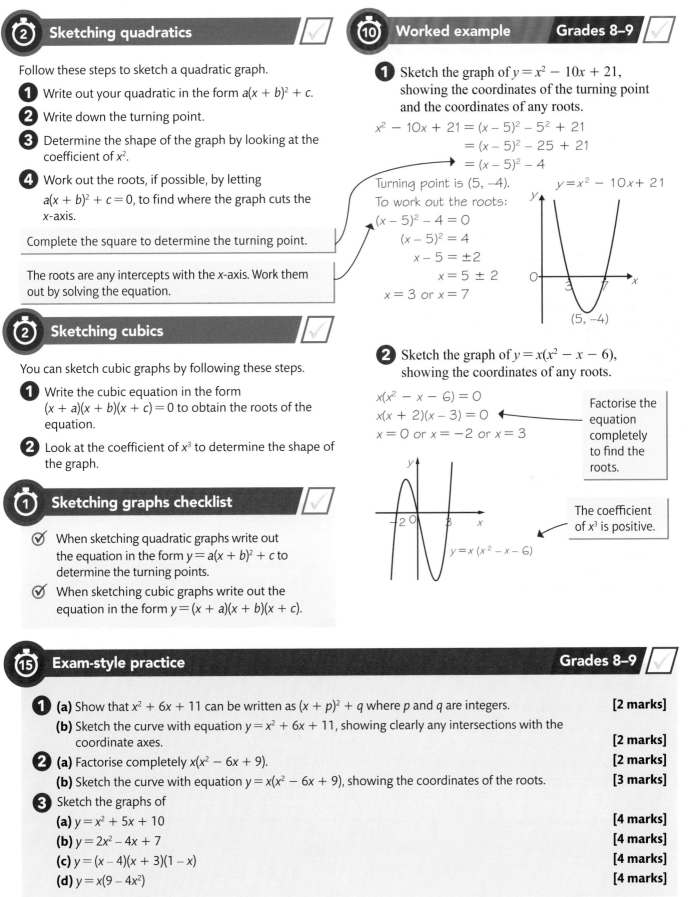

$y = x^2 - 10x + 21$

$(5, -4)$

2 Sketch the graph of $y = x(x^2 - x - 6)$, showing the coordinates of any roots.

$$x(x^2 - x - 6) = 0$$
$$x(x + 2)(x - 3) = 0$$
$$x = 0 \text{ or } x = -2 \text{ or } x = 3$$

> Factorise the equation completely to find the roots.

$y = x(x^2 - x - 6)$

> The coefficient of x^3 is positive.

⑮ Exam-style practice — Grades 8–9

1 (a) Show that $x^2 + 6x + 11$ can be written as $(x + p)^2 + q$ where p and q are integers. **[2 marks]**

(b) Sketch the curve with equation $y = x^2 + 6x + 11$, showing clearly any intersections with the coordinate axes. **[2 marks]**

2 (a) Factorise completely $x(x^2 - 6x + 9)$. **[2 marks]**

(b) Sketch the curve with equation $y = x(x^2 - 6x + 9)$, showing the coordinates of the roots. **[3 marks]**

3 Sketch the graphs of

(a) $y = x^2 + 5x + 10$ **[4 marks]**

(b) $y = 2x^2 - 4x + 7$ **[4 marks]**

(c) $y = (x - 4)(x + 3)(1 - x)$ **[4 marks]**

(d) $y = x(9 - 4x^2)$ **[4 marks]**

☑ Made a start   ☑ Feeling confident   ☑ Exam ready

Exponential graphs

An equation of the form $y = ka^x$ is called an exponential function. You need to be able to recognise and sketch exponential graphs. They typically appear in questions involving growth (compound interest, bacteria populations) and decay (nuclear half-life).

2 Shape of an exponential graph

An **exponential** graph has this shape.

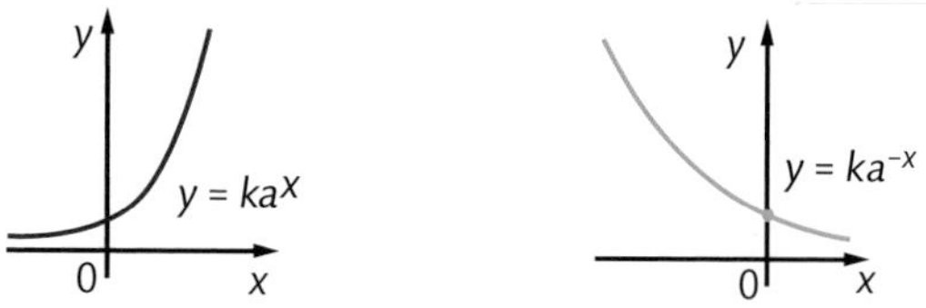

In both of these graphs the point of intersection on the y-axis is (0, k). Note that the graphs both approach the x-axis, but never reach it. This axis is an **asymptote** of the curve.

10 Worked example — Grades 5–6

1 (a) Complete the table of values for $y = 3^x$.

x	–1	0	1	2	3
y	0.33	1	3	9	27

(b) Draw the graph of $y = 3^x$ for values of x from -1 to 3.

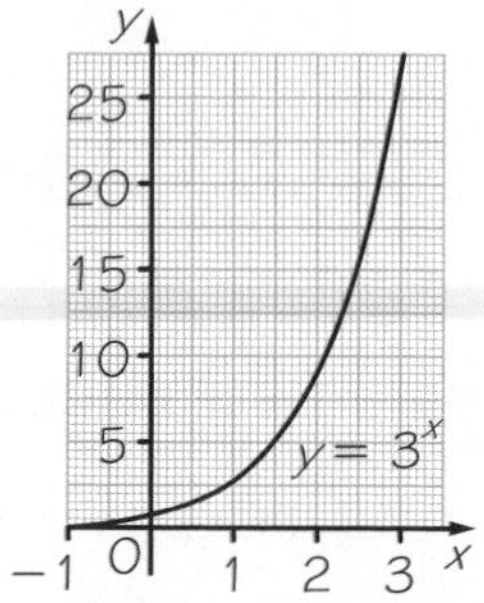

2 A scientist is growing a particular species of bacterium in the laboratory. Each bacterium divides in two every minute. He starts with one bacterium.

(a) How many bacteria are there after 5 minutes?

Number of bacteria $= 2 \times 2 \times 2 \times 2 \times 2$
$= 2^5 = 32$

(b) Write down an expression for the number of bacteria after t minutes.

Number of bacteria $= 2^t$

(c) Sketch a graph to show the number of bacteria after t minutes.

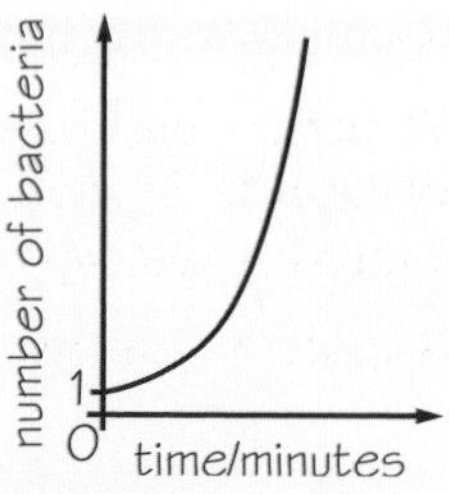

5 Worked example — Grade 5

A radioactive gas occurring naturally in rocks has a half-life of 1 minute. In a sample of rock, the gas was found to have 10 000 atoms.

Having a half-life of 1 minute means the sample will contain half as many atoms each time 1 minute passes.

(a) How many atoms will be present after 3 minutes?

After 1 minute: $10000 \div 2 = 5000$ atoms
After 2 minutes: $5000 \div 2 = 2500$ atoms
After 3 minutes: $2500 \div 2 = 1250$ atoms

(b) Write down a formula for the number of atoms after t minutes.

Number of atoms $= 10000 \times 0.5^t$

(c) Sketch a graph to show the number of atoms after t minutes.

Initially, number of atoms $= 10000 \times 0.5^0$
$= 10000 \times 1 = 10000$

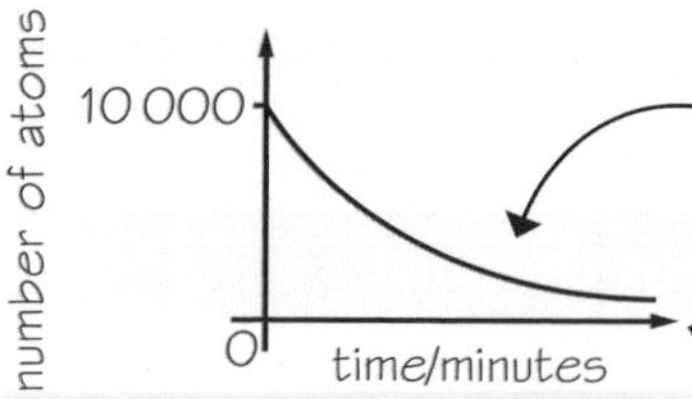

This is an example of exponential decay.

To sketch the graph you should plot a few initial values to get an idea of the shape of the graph.

10 Exam-style practice — Grades 5–9

1 Several years ago, a lake had a population of 200 000 fish. Overfishing has caused the population to decline by 15% each year.

(a) Show that the population of the fish can be modelled by $P = 200\,000 \times 0.85^t$ where t is the number of years. **[2 marks]**

(b) Work out the number of fish present after 10 years. **[2 marks]**

(c) Sketch a graph to show the number of fish after t years. **[2 marks]**

2 The equation of a curve has the form $y = ab^x$.
The curve passes through the points (1, 5) and (4, 320). Work out the value of a and the value of b. **[4 marks]**

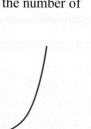

Gradients of curves

You need to be able to estimate the gradient of a curve at a given point.

⑤ Finding a gradient ☑

The gradient of a curve tells you the rate of change of the vertical variable with respect to the horizontal variable. The gradient of the curve at a point is the gradient of the **tangent** to the curve at that point. The tangent is the straight line that is parallel to the curve, at the point of contact.

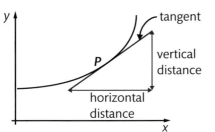

To work out the gradient at point P:

1 Lay a ruler along the curve at point P and draw a straight line to give the tangent at point P.

2 Draw a right-angled triangle using the tangent as the hypotenuse. Use the grid lines to give you sides with lengths you can work out easily.

3 Work out the gradient:
$$\text{gradient} = \frac{\text{change in vertical distance}}{\text{change in horizontal distance}}$$

⑤ Worked example Grade 9 ☑

In an experiment a liquid is heated to $70\,°C$ and then allowed to cool down. The graph shows this information.

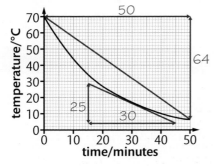

(a) Work out the average rate of decrease in temperature between $t = 0$ and $t = 50$.

$$\text{average rate of decrease} = \frac{64}{50}$$
$$= 1.28\,°C/\text{minute}$$

(b) Work out the rate of decrease in temperature at $t = 30$. State the units.

$$\text{rate of decrease} = \frac{20}{30} = 0.667\,°C/\text{minute}$$

⑤ Worked example Grade 9 ☑

Jake bought a house.

The graph shows the value of the house, in thousands of pounds, and the number of years.

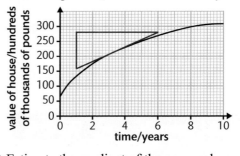

(a) Estimate the gradient of the curve when $t = 3$.

$$\text{gradient} = \frac{\text{change in vertical distance}}{\text{change in horizontal distance}}$$

$$\text{gradient} = \frac{280\,000 - 160\,000}{6 - 1} = \frac{120\,000}{5}$$
$$= 24\,000$$

(b) Interpret your answer to part **(a)**.

This shows that the rate of growth of the house value is £24 000 per year.

A positive gradient means that the rate of change is increasing and a negative gradient means that the rate of change is decreasing. See page 34 if you need to review gradients.

Draw a straight line joining the ends of the curve to find the average.

⑩ Exam-style practice Grade 9 ☑

A tank was emptied and the depth of water was recorded over ten minutes.

Time/ minutes	0	1	2	3	4	5	6	7	8	9	10
Depth/ metres	30	29	28	27	26	24	23	21.5	19	15	9

(a) Plot the graph of depth against time. **[3 marks]**
(b) Work out the average rate of decrease of the depth of the water between $t = 0$ and $t = 10$. **[2 marks]**
(c) Work out the gradient at $t = 7$. **[2 marks]**
(d) Give an interpretation of your answer to part **(c)**. **[1 mark]**

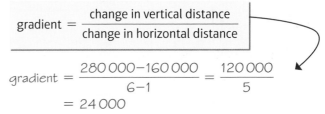

☑ **Made a start** ☑ **Feeling confident** ☑ **Exam ready**

BBC

Velocity–time graphs

A velocity–time graph shows how velocity changes with time over a journey. Velocity means speed in a given direction.

5 Velocity–time graphs

You must be able to work with and interpret velocity–time graphs. There are two very important facts about these graphs.

See page 41 to review how to draw a tangent to find the gradient of a curve.

If acceleration is negative then it is called deceleration.

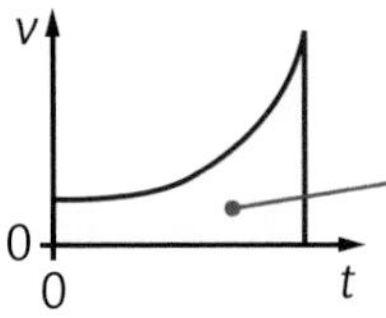

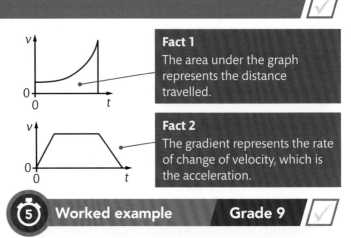

Fact 1
The area under the graph represents the distance travelled.

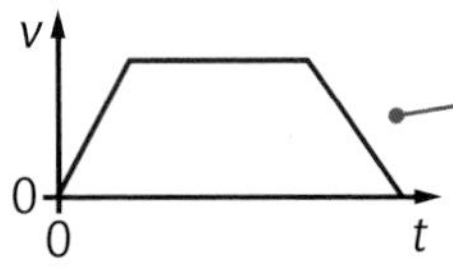

Fact 2
The gradient represents the rate of change of velocity, which is the acceleration.

5 Worked example Grade 8

The velocity–time graph shows the velocity of an object for the first 40 seconds of a journey.

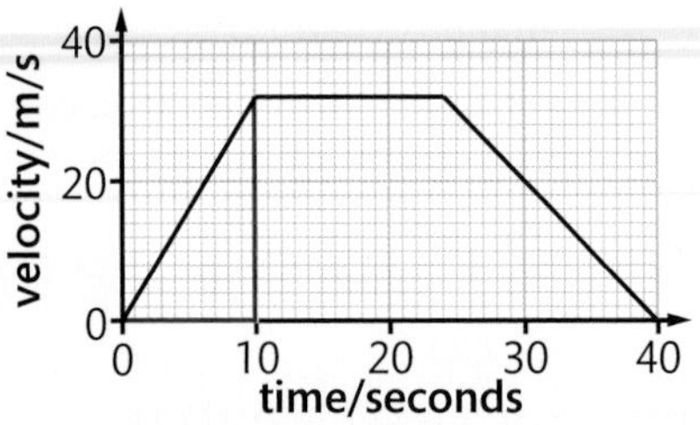

(a) (i) Work out the gradient for the first 10 seconds.

gradient $= \dfrac{32}{10} = 3.2$

(ii) Interpret your result.

The object is accelerating at $3.2\,\text{m/s}^2$.

(b) Work out the total distance travelled by the object in 40 seconds.

Area $= \dfrac{1}{2}(10 \times 32) + (14 \times 32) + \dfrac{1}{2}(16 \times 32)$

$= 160 + 448 + 256$

$= 864$

Distance travelled $= 864\,\text{m}$

Divide the region under the graph into two triangles and a rectangle.

You could alternatively use the formula for the area of a trapezium $A = \frac{1}{2}(a + b)h$.

5 Worked example Grade 9

The graph shows the velocity of a model rocket for the first 40 seconds of its flight.

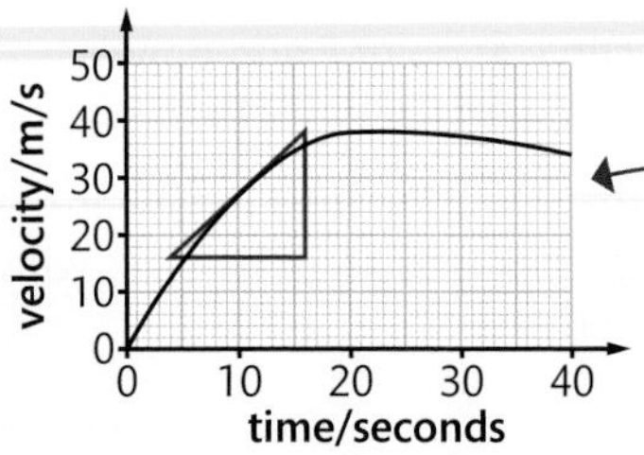

The tangent to a curved velocity–time graph allows you to work out the gradient.

(a) Estimate the gradient at $t = 10$ seconds.

gradient $= \dfrac{22}{12} = 1.83$

(b) Interpret your result from part **(a)**.

The rocket is accelerating at $1.83\,\text{m/s}^2$.

1 Checklist

- The gradient of a velocity–time graph gives the acceleration.
- The area under a velocity–time graph gives the distance travelled.

10 Exam-style practice Grades 8–9

A toy car is resting on a flat concrete floor. The toy car moves in a straight line, starting from rest. It travels with a constant acceleration for the first 5 seconds. It reaches a velocity of 8 m/s. It then travels at this velocity for a further 10 seconds. It then slows down and stops. The total time taken for the journey is 24 seconds.

(a) Draw a velocity–time graph for the toy car. **[3 marks]**

(b) Work out the total distance travelled by the toy car for the first 24 seconds. **[3 marks]**

Areas under curves

The area under a curve can be estimated by splitting the area into trapezia. For some graphs, the area under the curve has a specific meaning, for example, in velocity–time graphs, it gives the distance travelled.

5 Estimating areas

To estimate the area under a curve, split the region you require into simple shapes such as rectangles, triangles or trapezia. Then calculate the area of each shape, and add them up. For example, to work out the area under this curve, split it into trapezia.

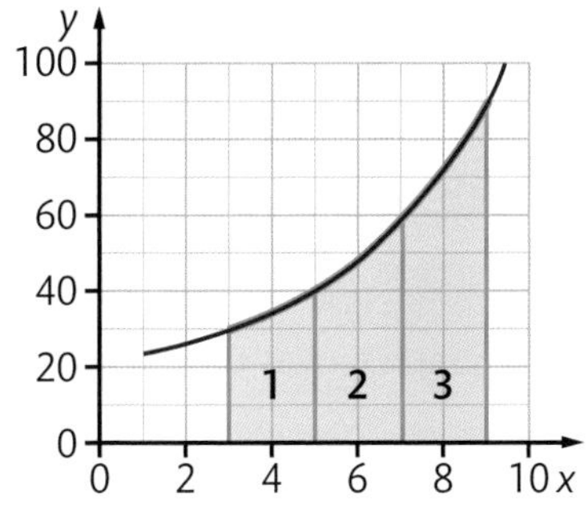

Area under curve = Area 1 + Area 2 + Area 3

$= [\frac{1}{2}(30 + 40) \times 2] + [\frac{1}{2}(40 + 60) \times 2] + [\frac{1}{2}(60 + 90) \times 2]$

$= 70 + 100 + 150$

$= 320$ units^2

In this case, the area of each trapezium is more than the area of each shaded section so it an **overestimate**.

As you increase the number of strips, the area gets nearer to its actual value.

Make sure you know how to work out the area of a trapezium.

5 Worked example — Grade 9

The graph shows the first 10 seconds of the journey of a ball rolling down a hill.

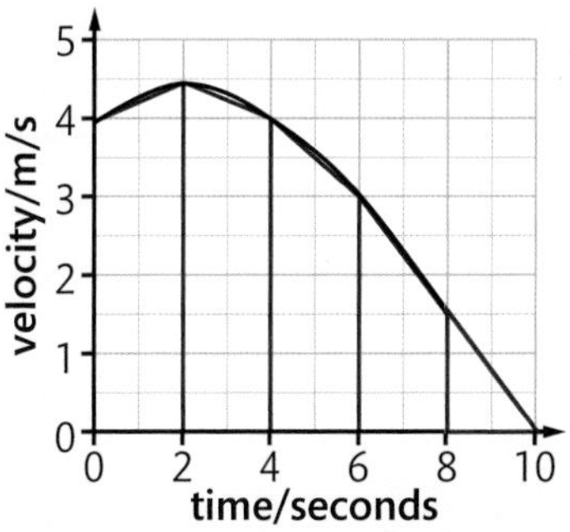

(a) Use the graph to estimate the total distance travelled in the first 10 seconds.

Area =

$[\frac{1}{2}(4 + 4.5) \times 2] + [\frac{1}{2}(4.5 + 4) \times 2] + [\frac{1}{2}(4 + 3) \times 2] + [\frac{1}{2}(3 + 1.5) \times 2] + [\frac{1}{2} \times 1.5 \times 2]$

$= 8.5 + 8.5 + 7 + 4.5 + 1.5 = 30$ m

(b) Is the answer to part **(a)** an overestimate or an underestimate? Justify your answer.

It is an underestimate, because the area of each trapezium and the triangle is below the curve.

2 Checklist

- Split the area into equal intervals.
- Work out the area of each trapezium and then add them together.
- If the trapezia are below the curve then it is an underestimate.
- If the trapezia are above the curve then it is an overestimate.

Exam focus

In the exam you can draw on the graphs to help you answer. Here, you would draw vertical lines to divide the area into equal strips.

10 Exam-style practice — Grade 9

This is a velocity–time graph for a skier.

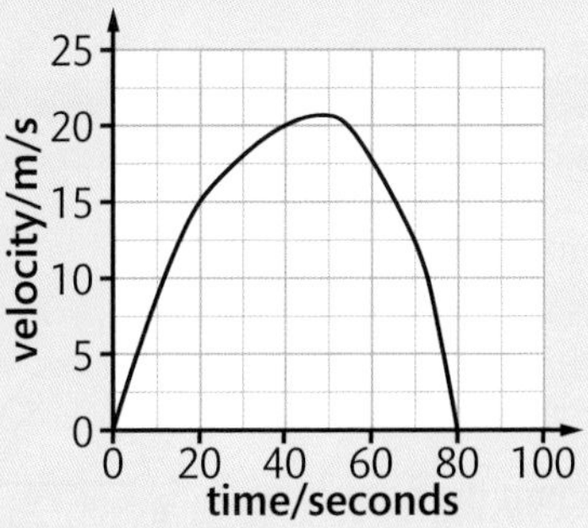

(a) Use the graph to estimate the total distance travelled in the first 80 seconds. **[4 marks]**

(b) Explain whether the answer to part **(a)** is an overestimate or an underestimate. **[1 mark]**

Made a start | Feeling confident | Exam ready

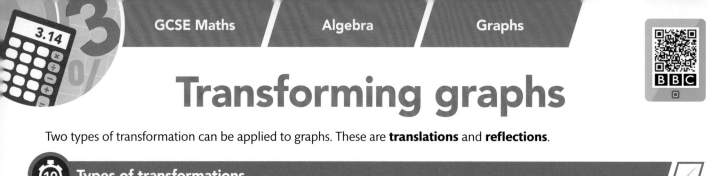

Transforming graphs

Two types of transformation can be applied to graphs. These are **translations** and **reflections**.

🕙 Types of transformations

These vectors show $\left(\dfrac{movement\ in\ x\text{-}direction}{movement\ in\ y\text{-}direction}\right)$

Transformation	$f(x) + a$	$f(x) - a$
Description (vector notation)	Translation $\begin{pmatrix} 0 \\ a \end{pmatrix}$	Translation $\begin{pmatrix} 0 \\ -a \end{pmatrix}$
What really happens?	$y = f(x)$ → $y = f(x) + a$. The graph moves **up** by a units.	$y = f(x)$, $y = f(x) - a$. The graph moves **down** by a units.

Transformation	$f(x + a)$	$f(x - a)$
Description	Translation $\begin{pmatrix} -a \\ 0 \end{pmatrix}$	Translation $\begin{pmatrix} a \\ 0 \end{pmatrix}$
What really happens?	$y = f(x + a)$, $y = f(x)$. The graph moves **to the left** by a units.	$y = f(x)$, $y = f(x - a)$. The graph moves **to the right** by a units.

Transformation	$-f(x)$	$f(-x)$
Description	Reflection in x-axis	Reflection in y-axis
What really happens?	The graph is reflected in the **x-axis**. $y = f(x)$, $y = -f(x)$	The graph is reflected in the **y-axis**. $y = f(x)$, $y = f(-x)$

🕔 Worked example   Grade 9

The diagram shows part of the curve $y = f(x)$. The coordinates of the maximum point of the curve are $(2, 5)$.

$(2, 5)$ $y = -f(x)$

$(2, -5)$ $y = f(x)$

(a) Write down the coordinates of the maximum point of the curve with equation

(i) $y = f(x + 3)$ (ii) $y = f(x) + 3$

$(-1, 5)$ $(2, 8)$

(b) Sketch the graph of $y = -f(x)$. Describe the type of transformation.

Reflection in the x-axis.

🕙 Exam-style practice   Grade 9

The diagram shows part of the curve $y = f(x)$. The coordinates of the minimum point of the curve are $(3, -4)$.

$(3, -4)$

(a) Write down the coordinates of the turning point of the curve with equation

(i) $y = f(x + 2)$ **[1 mark]**
(ii) $y = -f(x)$ **[1 mark]**

The curve with equation $y = f(x)$ is transformed to give the curve with equation $y = f(x) - 5$.

(b) Describe the transformation.   **[2 marks]**

✓ Made a start   ✓ Feeling confident   ✓ Exam ready

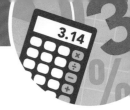

Algebraic fractions

Algebraic fractions are simply fractions with algebraic expressions in the numerator and/or denominator. To review work on fractions, see pages 1, 2 and 18.

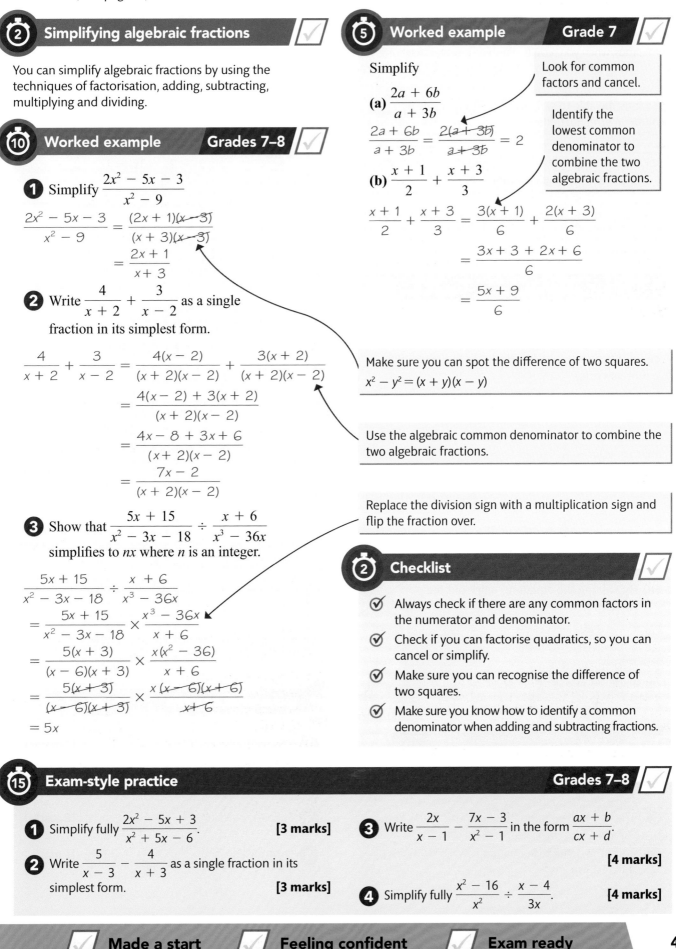

2 Simplifying algebraic fractions

You can simplify algebraic fractions by using the techniques of factorisation, adding, subtracting, multiplying and dividing.

10 Worked example — Grades 7–8

1 Simplify $\frac{2x^2 - 5x - 3}{x^2 - 9}$

$$\frac{2x^2 - 5x - 3}{x^2 - 9} = \frac{(2x + 1)\cancel{(x - 3)}}{(x + 3)\cancel{(x - 3)}}$$
$$= \frac{2x + 1}{x + 3}$$

Make sure you can spot the difference of two squares. $x^2 - y^2 = (x + y)(x - y)$

2 Write $\frac{4}{x + 2} + \frac{3}{x - 2}$ as a single fraction in its simplest form.

$$\frac{4}{x + 2} + \frac{3}{x - 2} = \frac{4(x - 2)}{(x + 2)(x - 2)} + \frac{3(x + 2)}{(x + 2)(x - 2)}$$
$$= \frac{4(x - 2) + 3(x + 2)}{(x + 2)(x - 2)}$$
$$= \frac{4x - 8 + 3x + 6}{(x + 2)(x - 2)}$$
$$= \frac{7x - 2}{(x + 2)(x - 2)}$$

Use the algebraic common denominator to combine the two algebraic fractions.

3 Show that $\frac{5x + 15}{x^2 - 3x - 18} \div \frac{x + 6}{x^3 - 36x}$ simplifies to nx where n is an integer.

$$\frac{5x + 15}{x^2 - 3x - 18} \div \frac{x + 6}{x^3 - 36x}$$
$$= \frac{5x + 15}{x^2 - 3x - 18} \times \frac{x^3 - 36x}{x + 6}$$
$$= \frac{5(x + 3)}{(x - 6)(x + 3)} \times \frac{x(x^2 - 36)}{x + 6}$$
$$= \frac{5\cancel{(x + 3)}}{\cancel{(x - 6)}\cancel{(x + 3)}} \times \frac{x\cancel{(x - 6)}\cancel{(x + 6)}}{\cancel{x + 6}}$$
$$= 5x$$

Replace the division sign with a multiplication sign and flip the fraction over.

5 Worked example — Grade 7

Simplify

(a) $\frac{2a + 6b}{a + 3b}$

$$\frac{2a + 6b}{a + 3b} = \frac{2\cancel{(a + 3b)}}{\cancel{a + 3b}} = 2$$

Look for common factors and cancel.

(b) $\frac{x + 1}{2} + \frac{x + 3}{3}$

$$\frac{x + 1}{2} + \frac{x + 3}{3} = \frac{3(x + 1)}{6} + \frac{2(x + 3)}{6}$$
$$= \frac{3x + 3 + 2x + 6}{6}$$
$$= \frac{5x + 9}{6}$$

Identify the lowest common denominator to combine the two algebraic fractions.

2 Checklist

- ✓ Always check if there are any common factors in the numerator and denominator.
- ✓ Check if you can factorise quadratics, so you can cancel or simplify.
- ✓ Make sure you can recognise the difference of two squares.
- ✓ Make sure you know how to identify a common denominator when adding and subtracting fractions.

15 Exam-style practice — Grades 7–8

1 Simplify fully $\frac{2x^2 - 5x + 3}{x^2 + 5x - 6}$. **[3 marks]**

2 Write $\frac{5}{x - 3} - \frac{4}{x + 3}$ as a single fraction in its simplest form. **[3 marks]**

3 Write $\frac{2x}{x - 1} - \frac{7x - 3}{x^2 - 1}$ in the form $\frac{ax + b}{cx + d}$. **[4 marks]**

4 Simplify fully $\frac{x^2 - 16}{x^2} \div \frac{x - 4}{3x}$. **[4 marks]**

Made a start | Feeling confident | Exam ready

Quadratics and fractions

When some equations containing algebraic fractions are simplified, they can be expressed as quadratic equations.

② Simplifying

Sometimes, when you have to combine algebraic fractions, your answer will simplify to produce a quadratic expression or equation which can be written in the form $ax^2 + bx + c = 0$. Then you can solve the quadratic by one of the methods on pages 21 to 23.

You must be able to simplify the left-hand side by combining the fractions.

Once you have simplified the algebraic fractions, cross multiply to remove the denominators.

Simplify to produce a quadratic and then solve the equation.

⑤ Worked example   Grade 9

A bag of 40 sweets can be shared equally between n people. If the same bag of sweets is shared equally between $(n - 1)$ people, each person gets 2 more sweets.

(a) Show that $n^2 - n - 20 = 0$.

$$\frac{40}{n - 1} - \frac{40}{n} = 2$$

$$\frac{40n - 40(n - 1)}{n(n - 1)} = 2$$

$$\frac{40}{n(n - 1)} = 2$$

Identify a common denominator and simplify the LHS.

$$40 = 2n(n - 1)$$
$$2n^2 - 2n - 40 = 0$$
$$n^2 - n - 20 = 0$$

(b) By solving the equation, work out the value of n.

$$n^2 - n - 20 = 0$$
$$(n - 5)(n + 4) = 0$$
$$n = 5 \text{ or } n = -4$$

n cannot be negative so solution is $n = 5$.

② Checklist

- ☑ Simplify the algebraic fractions (as shown on page 45) and multiply out brackets.
- ☑ The aim is to write out the quadratic in the form $ax^2 + bx + c = 0$.
- ☑ Solve the quadratic equation.

⑤ Worked example   Grade 9

Solve the equation $\dfrac{3}{x + 2} - \dfrac{1}{3x - 1} = \dfrac{1}{2}$.

$$\frac{3(3x - 1)}{(x + 2)(3x - 1)} - \frac{1(x + 2)}{(x + 2)(3x - 1)} = \frac{1}{2}$$

$$\frac{3(3x - 1) - 1(x + 2)}{(x + 2)(3x - 1)} = \frac{1}{2}$$

$$\frac{9x - 3 - x - 2}{(x + 2)(3x - 1)} = \frac{1}{2}$$

$$\frac{8x - 5}{(x + 2)(3x - 1)} = \frac{1}{2}$$

$$2(8x - 5) = (x + 2)(3x - 1)$$
$$16x - 10 = 3x^2 + 5x - 2$$
$$3x^2 - 11x + 8 = 0$$
$$(3x - 8)(x - 1) = 0$$

$$x = \frac{8}{3} \text{ or } x = 1$$

$$ax^2 + bx + c = 0$$

Problem solving

Use the information to make an equation. Then rearrange your equation to get the equation shown in the question. The question says 'Show that', so you need to show every step of your working.

The 'difference' tells you that one term minus the other gives you 2.

Multiply through by the denominator on the LHS, rearrange and simplify the equation.

⑮ Exam-style practice   Grade 9

❶ Solve the equation $\dfrac{2}{x} - \dfrac{3}{x + 1} = 2$.   **[4 marks]**

❷ Solve the equation $\dfrac{2}{x + 2} - \dfrac{3}{x - 4} = 2$.   **[5 marks]**

❸ Solve the equation $\dfrac{3}{x + 1} + \dfrac{2}{x + 5} = 1$.   **[5 marks]**

❹ Solve the equation $\dfrac{1}{x + 1} = \dfrac{1}{2} + \dfrac{1}{x + 3}$.   **[5 marks]**

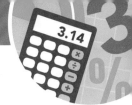

Function notation

You can use algebraic expressions to define functions.

2 Function notation

A **function** is a rule that describes one or more operations performed on a variable.

$f(x)$ means 'a function of x'.

This means that the variable in the algebraic expression is x.

For example, given the function $f(x) = x^2 + 1$,

$f(4)$ means 'the value of the function when x is replaced by 4'

so $f(4) = 4^2 + 1 = 17$

Substitute the given values.

Set the two functions equal and solve for x.

5 Composite functions

A composite function combines two or more functions in a single function. The output of one function is used as the input for the second function.

$f(x) = x^2 - 6$ and $g(x) = 2x + 9$

$gf(x) = 2(x^2 - 6) + 9$

$fg(x) = (2x + 9)^2 - 6$

gf(x) means substitute f(x) into g(x).

fg(x) means substitute g(x) into f(x).

2 Worked example — Grade 8

The functions f and g are defined as:

$f(x) = \dfrac{2}{x}$ and $g(x) = 4 - 3x$

Determine the following composite functions.

(a) $fg(x)$

$fg(x) = \dfrac{2}{4 - 3x}$

(b) $gf(x)$

$gf(x) = 4 - 3\left(\frac{2}{x}\right) = 4 - \frac{6}{x}$

1 Checklist

- To evaluate a function, substitute for the variable in the function.
- Always use the correct order of substitution when working with composite functions.
- Remember that $fg(x)$ is not generally equal to $gf(x)$.

10 Worked example — Grade 7

1 The functions f and g are defined for all real values of x such that $f(x) = x^2 - 6$ and $g(x) = 2x + 9$.

Work out

(a) $f(-4)$

$f(-4) = (-4)^2 - 6 = 16 - 6 = 10$

(b) $g(0.4)$

$g(0.4) = 2(0.4) + 9 = 0.8 + 9 = 9.8$

(c) For what two values of x does $f(x) = g(x)$?

$x^2 - 6 = 2x + 9$

$x^2 - 2x - 15 = 0$

$(x + 3)(x - 5) = 0$

$x = -3$ or $x = 5$

2 The functions f and g are defined by

$f(x) = 3x + 5$ and $g(x) = 4 - x$.

Solve the equation $gf(x) = 2x$.

$4 - (3x + 5) = 2x$

$4 - 3x - 5 = 2x$

$-1 = 5x$

$x = -\dfrac{1}{5}$

Substitute the given function f into g for x and solve for x.

When working with composite functions the order of substitution is very important.

If possible, always simplify the composite function to its simplest form.

10 Exam-style practice — Grades 7–8

1 The function f is defined as $f(x) = 2x - 3$.

(a) What is $ff(2)$? **[2 marks]**

(b) Solve the equation $ff(a) = a$. **[3 marks]**

2 The functions f and g are defined as

$f(x) = x^2$ and $g(x) = 5 + x$

(a) What is

(i) $fg(x)$

(ii) $gf(x)$? **[4 marks]**

(b) Show that there is a single value of x for which $fg(x) = gf(x)$ and determine this value of x. **[3 marks]**

Inverse functions

An inverse function reverses the operations of a function. The inverse function of $f(x) = x + 1$ is $f^{-1}(x) = x - 1$.

5 How to find an inverse function

For the function $f(x)$, the inverse function is $f^{-1}(x)$. You can use $f^{-1}(x)$ to find the value of the original variable x.

To find the inverse function of $f(x) = 3x - 5$:

1. Write as: $y = 3x - 5$ — Substitute f(x) for y and write the function as y = ...
2. Rearrange: $x = \frac{y + 5}{3}$ — Now rearrange it, to get x in terms of y.
3. Change x and y: $y = \frac{x + 5}{3}$ — Change x to y and change y to x.
4. Change y to $f^{-1}(x)$: $f^{-1}(x) = \frac{x + 5}{3}$ — Replace y with $f^{-1}(x)$. You now have the inverse function.

2 Worked example — Grade 9

Work out the inverse function of $f(x) = \frac{x + 3}{4}$.

$$y = \frac{x + 3}{4}$$
$$4y = x + 3$$
$$4y - 3 = x$$
$$y = 4x - 3$$
$$f^{-1}(x) = 4x - 3$$

Rearrange to get x in terms of y, then swap x and y.

Always write out the inverse function as $f^{-1}(x) = \ldots$

5 Worked example — Grade 9

The function f is such that $f(x) = \frac{8}{x + 2}$.

(a) What is $f^{-1}(x)$?

$$y = \frac{8}{x + 2}$$
$$y(x + 2) = 8$$
$$xy + 2y = 8$$
$$xy = 8 - 2y$$
$$x = \frac{8 - 2y}{y}$$
$$y = \frac{8 - 2x}{x}$$
$$f^{-1}(x) = \frac{8 - 2x}{x}$$

(b) Solve the equation $f^{-1}(x) = f(x)$.

$$\frac{8}{x + 2} = \frac{8 - 2x}{x}$$
$$8x = (8 - 2x)(x + 2)$$
$$8x = 16 + 8x - 2x^2 - 4x$$
$$2x^2 + 4x - 16 = 0$$
$$x^2 + 2x - 8 = 0$$
$$(x - 2)(x + 4) = 0$$
$$x = 2 \text{ or } x = -4$$

You might need to factorise to get x on its own when rearranging.

2 Checklist

- ✓ To find the inverse function swap the x and y.
- ✓ Rearrange the new equation for y.
- ✓ Write out the inverse function as $f^{-1}(x) = \ldots$

Problem solving

$ff^{-1}(x)$ is the composite function formed by working out $f^{-1}(x)$, then using this as the input for $f(x)$.

15 Exam-style practice — Grade 9

1. The function f is such that $f(x) = 7x - 3$.
   (a) What is $f^{-1}(x)$? **[2 marks]**
   (b) Solve the equation $f^{-1}(x) = f(x)$. **[3 marks]**

2. $f(x) = \frac{x}{x + 3}$, $x \neq -3$
   (a) Given that $f^{-1}(x) = -5$, what is the value of x? **[3 marks]**
   (b) Show that $ff^{-1}(x) = x$. **[3 marks]**

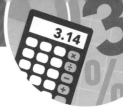

Equation of a circle

You need know how the equation of a circle relates to the radius of the circle.

2 Equation of a circle

Consider a circle, with centre at the origin 0 (0, 0) and radius r.

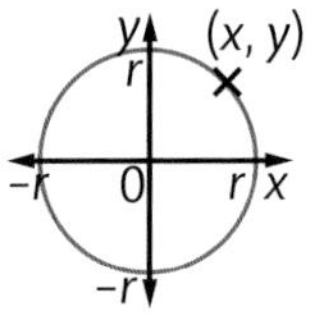

Then, using Pythagoras' theorem, for point (x, y) on the circle:

$x^2 + y^2 = r^2$

This is the equation of the circle.

2 Tangents to circles

When a line touches, but does not cross, the circumference of a circle it is called a **tangent** to the circle.

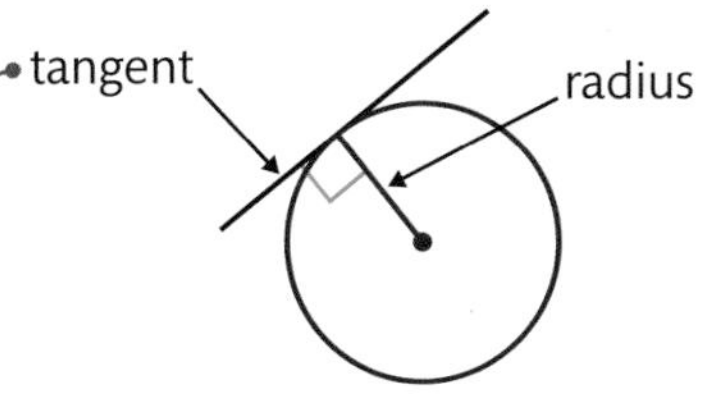

The tangent is perpendicular to the radius at the point where it touches the circumference.

The gradient of the line perpendicular to the radius is the negative reciprocal of the gradient of the radius.
$m_1 \times m_2 = -1$

1 Checklist

- The centre of the circle with equation $x^2 + y^2 = r^2$ is (0, 0) and r is the radius.
- A tangent is a line that just touches the circumference of a circle.

5 Worked example — Grade 8

C is the circle with equation $x^2 + y^2 = 40$.

(a) Show that the point $(-6, 2)$ lies on C.

$x^2 + y^2 = 40$
$(-6)^2 + (2)^2 = 40$
$36 + 4 = 40$
$40 = 40$

(b) Write down the radius of the circle. Give your answer in its simplest form.

$r^2 = 40$
$r = \sqrt{40}$
$= \sqrt{4}\sqrt{10}$
$= 2\sqrt{5}$

Be sure to simplify the surd into its simplest form.

5 Worked example — Grade 9

The equation of a circle is $x^2 + y^2 = 20$. The point $P(4, -2)$ lies on the circle. A tangent meets the circle at the point P. Work out the equation of the tangent.

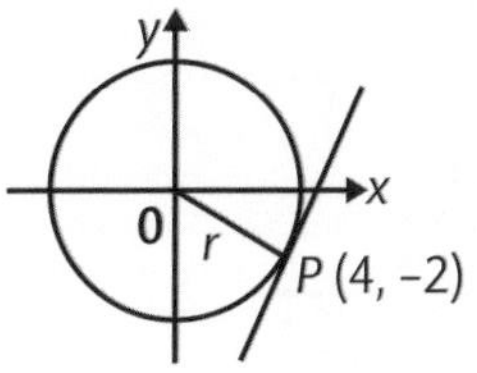

Sketch the graph to help visualise the problem.

gradient of the radius $= \dfrac{-2 - 0}{4 - 0} = -\dfrac{2}{4} = -\dfrac{1}{2}$

gradient of tangent $= 2$

$y = mx + c$
$-2 = 2(4) + c$
$-2 = 8 + c$
$-2 - 8 = c$
$c = -10$

Equation of tangent is $y = 2x - 10$.

10 Exam-style practice — Grade 9

1 **(a)** Solve the simultaneous equations $y = x - 4$ and $x^2 + y^2 = 8$. **[5 marks]**

(b) Describe the geometrical relationship between the line $y = x - 4$ and the circle $x^2 + y^2 = 8$ for this point of intersection. **[1 mark]**

2 The line L is a tangent to the circle $x^2 + y^2 = 10$ at the point A with coordinates (1, 3).

The line L crosses the x-axis at the point P.

Work out the area of triangle OAP. **[5 marks]**

Iteration

Iteration is a process you can use to estimate an answer to an equation. You use your first estimation to deduce a better estimate, then use this estimate to deduce an even better estimate, and so on.

(2) The iterative process ✓

Iteration is used to solve equations. The method involves repeating a process until you reach an accurate estimate for the solution. Each repetition of the process is also called an iteration.

The result of one iteration is used as the starting point for the next iteration.

> u_1 is the initial estimated value. Substitute this value into the iteration formula.

> Now substitute the value of 1.7321 into the iteration formula. Continue with this method to obtain the first four values.

(5) Worked example   Grade 9 ✓

Work out the first four terms of the sequence given by

$$x_{n+1} = \sqrt{1 + \frac{4}{x_n}} \text{ where } x_1 = 2$$

Give your answers to 4 decimal places.

$$x_2 = \sqrt{1 + \frac{4}{x_1}} = \sqrt{1 + \frac{4}{2}} = 1.7321$$

$$x_3 = \sqrt{1 + \frac{4}{x_2}} = \sqrt{1 + \frac{4}{1.7321}} = 1.8192$$

$$x_4 = \sqrt{1 + \frac{4}{x_3}} = \sqrt{1 + \frac{4}{1.8192}} = 1.7885$$

$$x_5 = \sqrt{1 + \frac{4}{x_4}} = \sqrt{1 + \frac{4}{1.7885}} = 1.7990$$

(5) Locating a root ✓

Suppose the graph of $y = f(x)$ is continuous and crosses the x-axis between $x = a$ and $x = b$.

The diagram shows this situation.

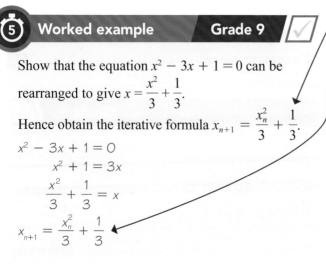

$f(a) < 0$ and $f(b) > 0$ $f(a) > 0$ and $f(b) < 0$

If there is a **change in sign** for y for two particular values of x then there is a **root** between these values of x. This means that the equation $f(x) = 0$ will have a solution between these two x-values.

(5) Worked example   Grade 9 ✓

Show that the equation $x^2 - 3x + 1 = 0$ can be

rearranged to give $x = \frac{x^2}{3} + \frac{1}{3}$.

Hence obtain the iterative formula $x_{n+1} = \frac{x_n^2}{3} + \frac{1}{3}$.

$x^2 - 3x + 1 = 0$

$x^2 + 1 = 3x$

$\dfrac{x^2}{3} + \dfrac{1}{3} = x$

$x_{n+1} = \dfrac{x_n^2}{3} + \dfrac{1}{3}$

(2) Worked example   Grade 9 ✓

Show that the equation $2 + 4x - x^4 = 0$ has a solution between $x = 1$ and $x = 2$.

$f(x) = 2 + 4x - x^4$

Given $x = 1$ and $x = 2$

$f(1) = 2 + 4(1) - 1^4 = 5 > 0$

$f(2) = 2 + 4(2) - 2^4 = -6 < 0$

As $f(1) > 0$ and $f(2) < 0$, there is a solution to $f(x) = 0$ between $x = 1$ and $x = 2$.

> To find the iteration formula from a quadratic equation, first rearrange the equation so that x is the subject.

> Then write out the iteration formula in terms of n.

(10) Exam-style practice   Grade 9 ✓

(a) Show that the equation $x^2 - 5x + 2 = 0$ has a root between $x = 4$ and $x = 5$.   **[2 marks]**

(b) Show that the equation $x^2 - 5x + 2 = 0$ can be arranged to give $x = \sqrt{5x - 2}$.   **[2 marks]**

(c) Use the iteration $x_{n+1} = \sqrt{5x_n - 2}$, with $x_1 = 5$, to work out a solution to the equation $x^2 - 5x + 2 = 0$ correct to 1 decimal place.   **[2 marks]**

Pages
13–50
LINKS

Algebra

Read the exam-style question and worked solution, then practise your exam skills with the question at the bottom of the page.

---

10 **Worked example**   Grade 9 ✓

(a) Write the equation $y = x^2 + 4x - 6$ in the form $y = (x + a)^2 + b$, where a and b are integers. Hence identify and write down the coordinates of the turning point of the curve.

$y = x^2 + 4x - 6$
$y = (x + 2)^2 - 4 - 6$
$y = (x + 2)^2 - 10$
The turning point occurs when
$(x + 2)^2 = 0$
$x + 2 = 0$
$x = -2$
When $x = -2$:
$y = (-2 + 2)^2 - 10 = 0 - 10 = -10$
Coordinates of the turning point are $(-2, -10)$.

(b) Sketch the curve.

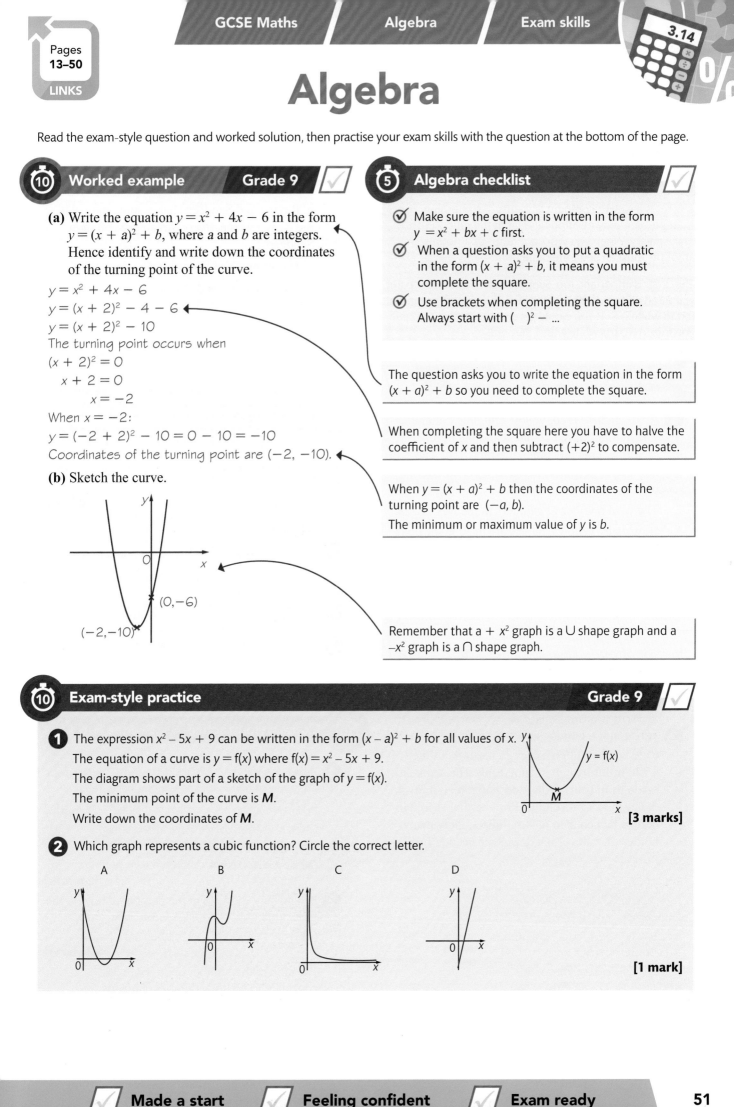

5 **Algebra checklist** ✓

☑ Make sure the equation is written in the form $y = x^2 + bx + c$ first.
☑ When a question asks you to put a quadratic in the form $(x + a)^2 + b$, it means you must complete the square.
☑ Use brackets when completing the square. Always start with ( )² – ...

The question asks you to write the equation in the form $(x + a)^2 + b$ so you need to complete the square.

When completing the square here you have to halve the coefficient of x and then subtract $(+2)^2$ to compensate.

When $y = (x + a)^2 + b$ then the coordinates of the turning point are $(-a, b)$.
The minimum or maximum value of y is b.

Remember that a $+ x^2$ graph is a ∪ shape graph and a $-x^2$ graph is a ∩ shape graph.

---

10 **Exam-style practice**   Grade 9 ✓

1 The expression $x^2 - 5x + 9$ can be written in the form $(x - a)^2 + b$ for all values of x.
The equation of a curve is $y = f(x)$ where $f(x) = x^2 - 5x + 9$.
The diagram shows part of a sketch of the graph of $y = f(x)$.
The minimum point of the curve is **M**.
Write down the coordinates of **M**. **[3 marks]**

2 Which graph represents a cubic function? Circle the correct letter.

A   B   C   D

[1 mark]

✓ **Made a start**   ✓ **Feeling confident**   ✓ **Exam ready**   **51**

Ratio

You can use ratios to describe how two or more quantities that are in proportion are related.

⑤ Ratio

A **ratio** is a way to compare amounts of something. For example, you may mix flour and fat in the ratio 2:1 to make pastry. When you work out ratios of amounts, the amounts must all be in the same unit.

The order is important, too. Suppose that to make pink paint, you mix 8 parts red with 1 part white. This means the ratio of red to white is 8:1. You would get a very different colour if you mixed 8 parts white with 1 part red.

⑩ Worked example   Grade 5

1 A box contains red, yellow and blue counters. The number of red counters and the number of yellow counters are in the ratio 7:4. The number of yellow counters and the number of blue counters are in the ratio 3:1.

The total number of counters in the box is 592.

How many yellow counters are there in the box?

red : yellow	yellow : blue
7 : 4	3 : 1
So: 21 : 12	12 : 4

Then: red : yellow : blue
           21 : 12 : 4

Number of yellow counters $= \dfrac{12}{37} \times 592 = 192$

Multiply each ratio by the yellow value in the other ratio.

2 160 children will be attending a school fete. Kate is going to give each of the 160 children a 240 millilitre cup of fruit drink. The drink is made from fruit cordial and water, mixed in the ratio 1:7 by volume. Fruit cordial is sold in bottles that contain 650 millilitres. How many bottles of fruit cordial does Kate need to buy?

Total = 1 + 7 = 8 parts
8 parts = 240 ml
1 part = 240 ÷ 8 = 30 ml
Amount of cordial per cup is 30 ml
Amount needed for all children is
160 × 30 = 4800 ml
Number of bottles of cordial is
4800 ÷ 650 = 7.384...
Number of bottles needed = 8

Seven bottles won't be enough, so round up.

⑤ Worked example   Grade 5

Anjali, Ravina and Sandeep shared some money in the ratio 7:5:3, respectively. Anjali got £60 more than Sandeep.

How much money did Ravina get?

Difference in Anjali's and Sandeep's amounts
= 7 parts – 3 parts = 4 parts
4 parts = £60
1 part = £60 ÷ 4 = £15
Ravina receives 5 × £15 = £75.

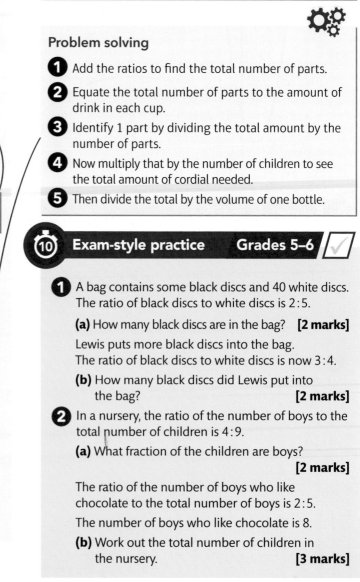

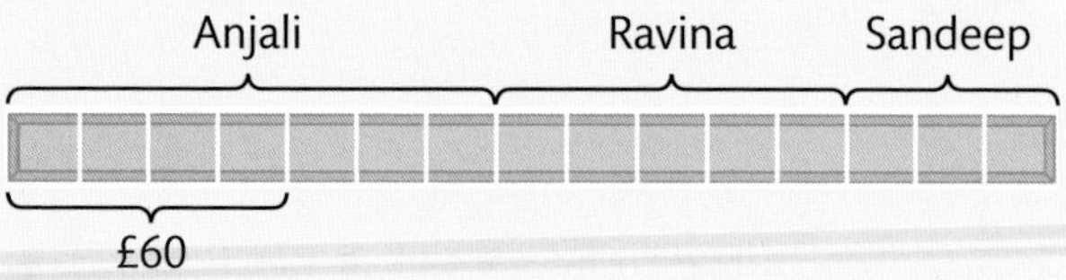

Anjali   Ravina   Sandeep

£60

Always work out one part when working with ratios.

Problem solving

1 Add the ratios to find the total number of parts.

2 Equate the total number of parts to the amount of drink in each cup.

3 Identify 1 part by dividing the total amount by the number of parts.

4 Now multiply that by the number of children to see the total amount of cordial needed.

5 Then divide the total by the volume of one bottle.

⑩ Exam-style practice   Grades 5–6

1 A bag contains some black discs and 40 white discs. The ratio of black discs to white discs is 2:5.

(a) How many black discs are in the bag? **[2 marks]**

Lewis puts more black discs into the bag.
The ratio of black discs to white discs is now 3:4.

(b) How many black discs did Lewis put into the bag? **[2 marks]**

2 In a nursery, the ratio of the number of boys to the total number of children is 4:9.

(a) What fraction of the children are boys? **[2 marks]**

The ratio of the number of boys who like chocolate to the total number of boys is 2:5.

The number of boys who like chocolate is 8.

(b) Work out the total number of children in the nursery. **[3 marks]**

✓ Made a start   ✓ Feeling confident   ✓ Exam ready

Proportion

Quantities may be related to each other by direct proportion or inverse proportion.

[2] Direct proportion

Two quantities are in **direct proportion** when both quantities increase or decrease at the same rate. For example:

1 kg of apples costs £1.05

× 8 × 8

8 kg of the same apples cost £8.40

[2] Worked example Grade 5

8 chopping boards cost £62.80. Work out the cost of 14 of these chopping boards.

Cost of 8 boards = £62.80

Cost of 1 board = £7.85

Cost of 14 boards is 14 × £7.85 = £109.90

You know the cost of 8 boards, so divide by 8 to give the cost of 1 board, then multiply this by 14.

÷ 8	8 boards	£62.80	÷ 8
	1 board	£7.85	
× 14	14 boards	£109.90	× 14

[10] Worked example Grade 5

1 A recruitment agency says that it takes 18 workers 8 days to put down a concrete base. A construction company wants to put down a concrete base in 3 days. How many workers do they need to hire from the recruitment agency?

8 days = 18 workers

1 day = 18 × 8 = 144 workers

3 days = 144 ÷ 3 = 48 workers

Remember that more days means fewer workers are needed and fewer days means more workers are needed.

2 Asha made some ginger biscuits to sell at a charity event. She had 6 kg of flour, 3.75 kg of butter, 2.4 kg of sugar and 355 g of ginger.

Recipe
Ginger biscuits
Makes 18
125 g flour
100 g butter
80 g sugar
10 g ginger

She made as many batches of 18 biscuits as she could, using the ingredients she had. How many batches of biscuits did she make?

Flour	6000 ÷ 125	= 48
Butter	3750 ÷ 100	= 37.5
Sugar	2400 ÷ 80	= 30
Ginger	355 ÷ 10	= 35.5

The number of batches of ginger biscuits is 30.

The lowest value gives the maximum possible number of batches.

Problem solving

Divide the amount of each ingredient by the amount needed in the recipe. Remember to convert all the ingredients to grams so that the ratios are all in the same unit.

[5] Inverse proportion

Two quantities are in **inverse proportion** when one quantity increases as the other quantity decreases, or vice versa.

Suppose it takes 6 workers 15 days to complete a job. How long will it take for 1 worker to complete the job? Think about how many 'worker days' the job takes.

1 worker takes 15 × 6 = 90 days

How long does it take for 2 workers to complete the job?

2 workers take 90 ÷ 2 = 45 days to complete the same job.

[1] Checklist

- ✓ For direct proportion, as one quantity increases the second quantity increases at the same rate, and vice versa.
- ✓ For inverse proportion, as one quantity increases the second quantity decreases, and vice versa.

[10] Exam-style practice Grades 5–6

1 Tom has a plank of wood. The wood has a length of 12 metres. Tom cuts the wood into two lengths, length *X* and length *Y*. The length *X* is 7 metres. The weight of length *X* is 10.5 kg. Work out the weight of length *Y*. **[3 marks]**

2 In the UK, diesel costs £1.19 per litre. In the USA, diesel costs 2.95 dollars per US gallon. Using 1 US gallon = 3.8 litres and £1 = 1.29 dollars, determine whether diesel is more expensive in the UK or in the USA. **[4 marks]**

3 An army base had provisions for 300 soldiers for 90 days. After 20 days, 50 soldiers left the army base. How long would the food last at the same rate? **[4 marks]**

Compound measures

A **compound measure** is made up of two or more other measurements.

2 Compound measures

A **compound measure** is a mathematical or scientific measurement made up of two or more other measurements.

Speed is a compound measure because it is made up of a time measurement and a distance measurement. You should be able to use standard and compound measures with numbers and algebra.

Problem solving

Convert 350 g/s into 0.35 kg/s because the units of any variable must be the same before you can make comparisons.

Write down a formula for the situation.

Then substitute into the formula to find each mass.

5 Worked example — Grade 5

Sand was falling from lorry A at a rate of 0.3 kg/s and it took 15 minutes for all the sand to fall.

Sand was falling from lorry B at a rate of 350 g/s and it took 12 minutes for all the sand to fall.

Which lorry was carrying more sand?

350 g ÷ 1000 = 0.35 kg

15 minutes = 900 s

12 minutes = 720 s

$\text{rate} = \frac{\text{mass}}{\text{time}}$

Lorry A:

$0.3 = \frac{\text{mass}}{900}$

mass = 0.3 × 900

= 270 kg

Lorry B:

$0.35 = \frac{\text{mass}}{720}$

mass = 0.35 × 720

= 252 kg

Lorry A was carrying more sand, as 270 kg is more than 252 kg.

10 Worked example — Grade 5

1 $\text{Pressure} = \frac{\text{force}}{\text{area}}$

Work out the pressure exerted by a force of 1200 N on a rectangular metal plate measuring 80 cm by 48 cm. Give your answer in N/m².

Area = 0.8 m × 0.48 m = 0.384 m²

The answer is to be in newtons per square metre, so you need to convert centimetres into metres.

$\text{Pressure} = \frac{\text{force}}{\text{area}} = \frac{1200}{0.384}$

= 3125 N/m²

2 A full tank contains 540 litres of water. When a tap is opened, water flows out at the rate of 0.4 litres per second. How long will it take to empty the tank? Give your answer in minutes.

$\text{Rate of flow} = \frac{\text{volume}}{\text{time}}$

$0.4 = \frac{540}{t} \Rightarrow t = \frac{540}{0.4}$

$t = 1350$ seconds

$t = 22.5$ minutes

You can work out the formula from the units. In this case, 'litres per second' means volume divided by time.

2 Checklist

- ✓ If you forget a formula you may be able to work it out from the given units. For example, if a density is given in g/cm³, you can see that density must be $\frac{\text{mass}}{\text{volume}}$.
- ✓ Make sure you read the question carefully and convert any units correctly, if necessary.

10 Exam-style practice — Grades 5–6

1 The force exerted by a raft on water is 1023 N. It exerts a pressure of 310 N/m² on the water.

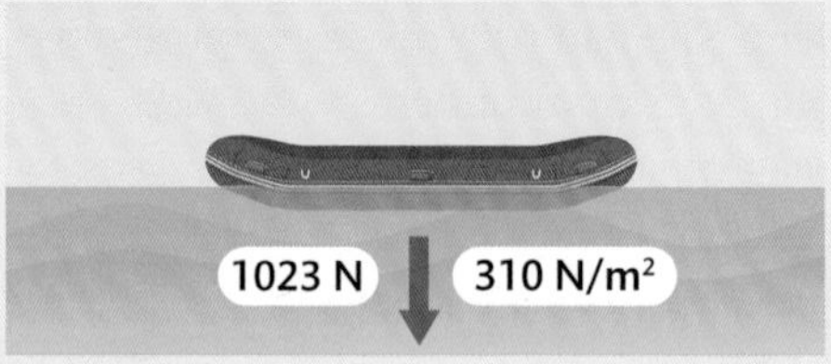

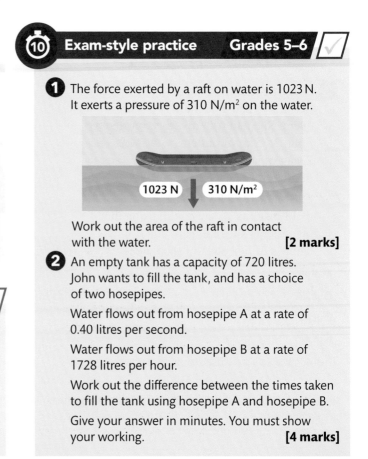

Work out the area of the raft in contact with the water. **[2 marks]**

2 An empty tank has a capacity of 720 litres. John wants to fill the tank, and has a choice of two hosepipes.

Water flows out from hosepipe A at a rate of 0.40 litres per second.

Water flows out from hosepipe B at a rate of 1728 litres per hour.

Work out the difference between the times taken to fill the tank using hosepipe A and hosepipe B.

Give your answer in minutes. You must show your working. **[4 marks]**

Speed

Speed describes how fast an object is moving. Speed is a compound measure (see page 54), with units such as miles per hour and metres per second.

(5) Formula for speed xy^2

Speed is a rate that compares distance with time.

The formula for speed is:

$$speed = \frac{distance\ travelled}{time\ taken}\ \text{or, more simply,}\ \frac{distance}{time}$$

This can be rearranged to:

$$time = \frac{distance}{speed}\ \text{and distance} = speed \times time$$

The speed formula can be summarised by this triangle. If you cover the variable you need to find, the visible part of the triangle shows you how to calculate it.

(5) Worked example — Grade 5

An athlete runs a distance of 660 metres in 1 minute and 15 seconds. Work out her average speed in metres per second.

0:01:15
MIN SEC

Total time is
60 seconds + 15 seconds = 75 seconds

$$speed = \frac{distance}{time} = \frac{660}{75} = 8.8\,m/s$$

Problem solving

1 You need to work out the distance he travelled in total. Make sure you divide the 30 minutes by 60, to convert to hours.

2 Rearrange the formula for distance to get time.

3 You need to work out the total time for which he travelled.

4 Now use the formula for speed.

(5) Worked example — Grade 6

On a journey from Paris to Lyon, René drove at an average speed of 50 mph for the first 2 hours and 30 minutes of his journey.

The remaining 210 miles of his journey were completed at an average speed of 70 mph.

René worked out that he had completed the whole journey at an average speed of 55 mph. Is René correct?

First part of the journey:
distance = speed × time = 50 × 2.5 = 125 miles
Total distance = 125 + 210 = 335 miles
Second part of the journey:

$$time = \frac{distance}{speed} = \frac{210}{70} = 3\ hours$$

Total time = 2.5 + 3 = 5.5 hours

$$Average\ speed = \frac{total\ distance}{total\ time}$$
$$= \frac{335}{5.5}$$
$$= 60.9\ mph\ (1\ d.p.)$$

No, he is not correct.

(10) Exam-style practice — Grades 5–6

1 Ted drove from Birmingham to Aberdeen. It took him 6 hours at an average speed of 70 mph. Jane drove from Birmingham to Aberdeen. She took 7 hours.
Work out Jane's average speed from Birmingham to Aberdeen. **[3 marks]**

2 Simon went for a bike ride one evening. He travelled x kilometres in 4 hours. Show that his average speed can be written as $\frac{5x}{72}$ m/s. **[3 marks]**

3 Angela is driving along a motorway in Europe to meet a friend. Her satnav is showing that she will reach the next turn-off, which is 40 miles away, in 31 minutes. The speed limit on the motorway is 80 mph. She thinks she will exceed the speed limit. Work out whether Angela is correct. You must show your working. **[3 marks]**

Density

Density is a compound measure (see page 54) that compares the mass of a given amount of a material to its volume. It is measured in units such as grams per cubic centimetre (g/cm³).

5 Formula for density xy^2

The formula for density is:

$$\text{density} = \frac{\text{mass}}{\text{volume}}$$

which can be rearranged to

$$\text{volume} = \frac{\text{mass}}{\text{density}} \text{ or mass} = \text{density} \times \text{volume}$$

The density formula can be summarised by this triangle. If you put the point of your pencil on the amount you need, the formula is along the opposite side.

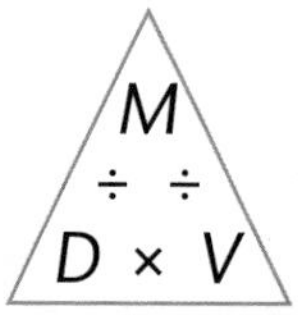

5 Worked example — Grade 5

210 g of zinc is mixed with 360 g of copper to make the alloy brass. The density of copper is 9 g/cm³. The density of zinc is 7 g/cm³.

(a) Work out the volume of zinc used in the alloy.

$$\text{Volume of zinc} = \frac{\text{mass}}{\text{density}} = \frac{210}{7} = 30\,\text{cm}^3$$

Use the version of the formula that will give the volume.

(b) What is the density of the alloy?

$$\text{Volume of copper} = \frac{\text{mass}}{\text{density}} = \frac{360}{9} = 40\,\text{cm}^3$$

Total volume of brass = 30 + 40 = 70 cm³

Total mass of brass = 210 + 360 = 570 g

$$\text{Density of brass} = \frac{\text{mass}}{\text{volume}} = \frac{570}{70} = 8.14\,\text{g/cm}^3$$

Use the total mass and the total volume to work out the density of the alloy.

Problem solving

When questions are given in context, you may need to identify which quantities you need to work out, before you start. Here, you need to work out the volume of the cuboid.

Volume = length × width × height

The formulae are given on page 72.

You need to change metres to centimetres because the density is given in g/cm³.

5 Worked example — Grade 5

A cuboid of rock measures 1 m by 0.7 m by 0.5 m. The density of the rock is 3.5 g/cm³. The rock needs to be placed on a pallet. Can the rock be supported by a pallet that holds a maximum mass of 1200 kg? Justify your decision.

Volume of cuboid = 100 cm × 70 cm × 50 cm
= 350 000 cm³

Mass of rock = density × volume
= 3.5 × 350 000
= 1 225 000 g
= 1225 kg

The rock cannot be supported by the pallet as 1225 kg is more than 1200 kg.

Convert grams to kilograms by dividing by 1000, to compare with the maximum mass that can be supported on the pallet.

10 Exam-style practice — Grades 5–6

1 **(a)** Work out the volume of the prism shown here. **[2 marks]**

The prism is made from wood and has a mass of 896 g.

(b) Work out the density, in g/cm³, of the wood. Give your answer correct to 3 significant figures. **[2 marks]**

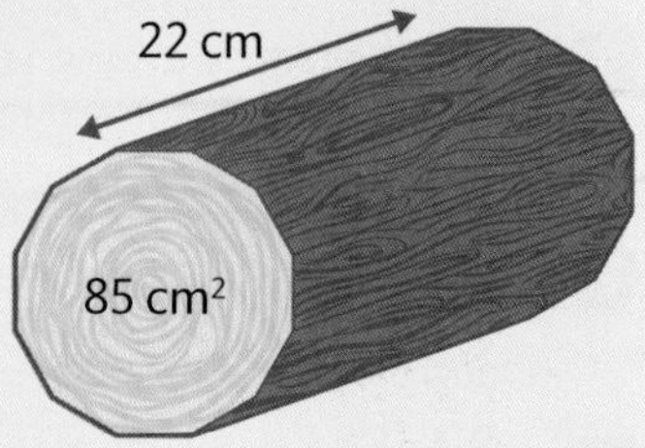

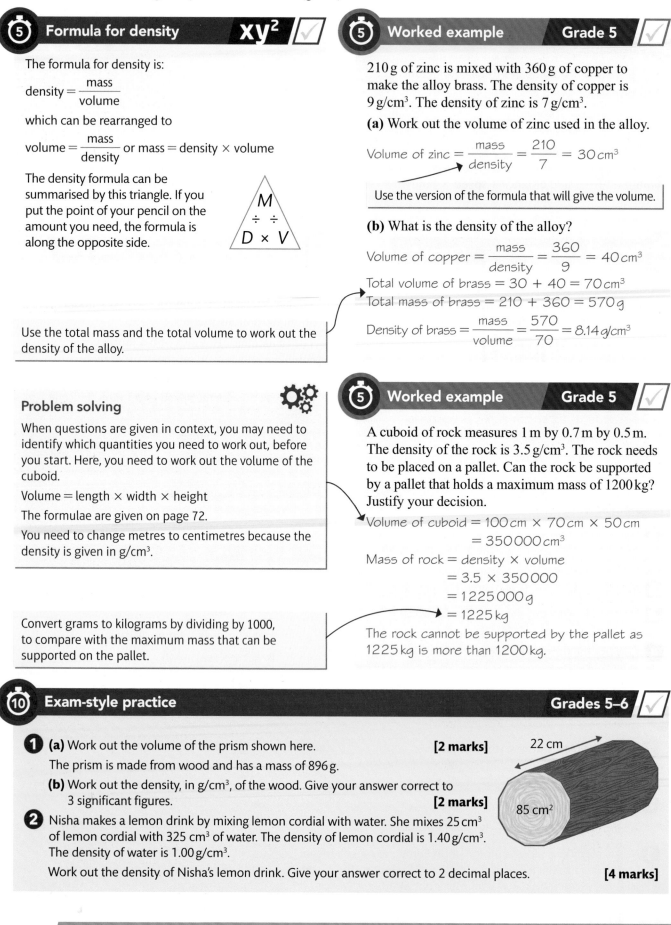

2 Nisha makes a lemon drink by mixing lemon cordial with water. She mixes 25 cm³ of lemon cordial with 325 cm³ of water. The density of lemon cordial is 1.40 g/cm³. The density of water is 1.00 g/cm³.

Work out the density of Nisha's lemon drink. Give your answer correct to 2 decimal places. **[4 marks]**

Proportion and graphs

Direct proportion and inverse proportion (see page 53) can be shown graphically.

5 Direct and inverse proportion

Direct proportion graphs look like this:

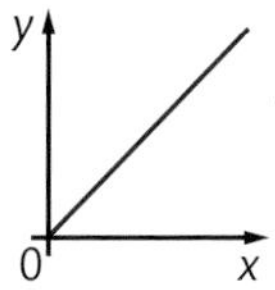

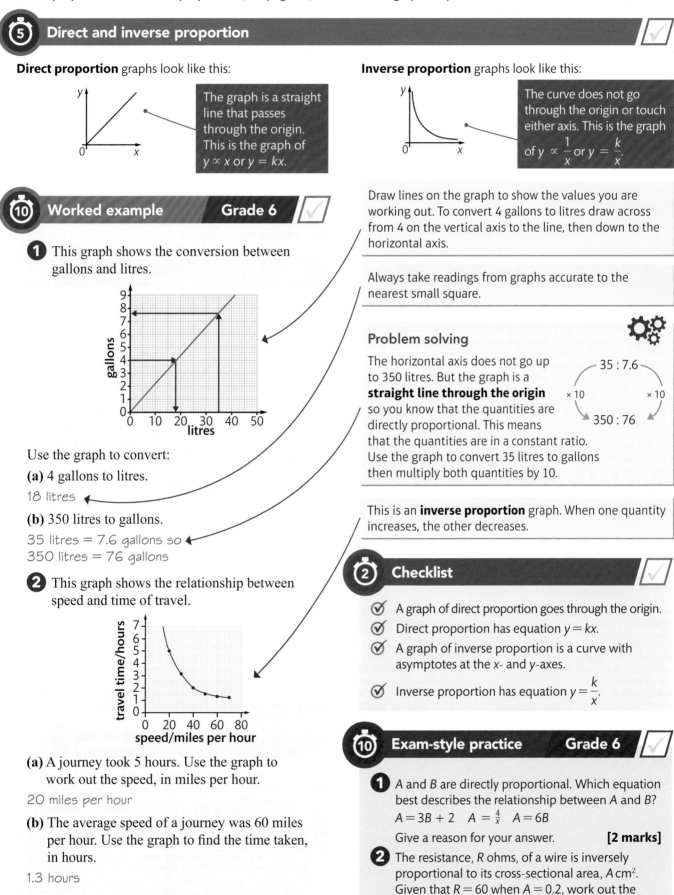

The graph is a straight line that passes through the origin. This is the graph of $y \propto x$ or $y = kx$.

Inverse proportion graphs look like this:

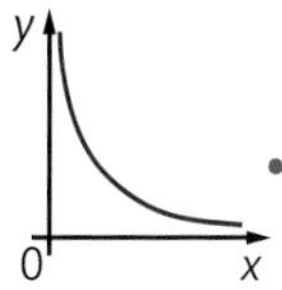

The curve does not go through the origin or touch either axis. This is the graph of $y \propto \frac{1}{x}$ or $y = \frac{k}{x}$.

10 Worked example — Grade 6

1 This graph shows the conversion between gallons and litres.

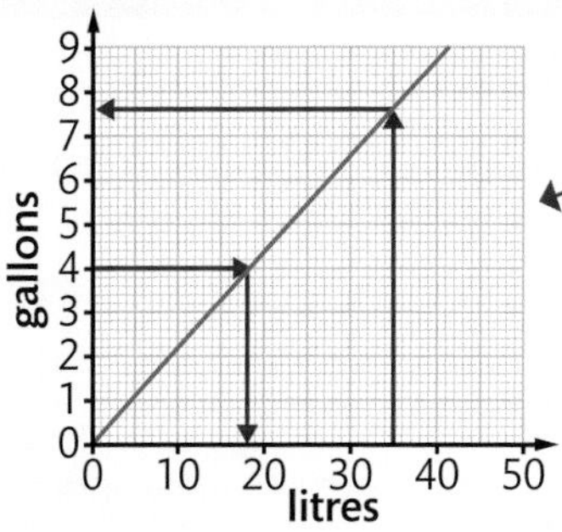

Use the graph to convert:

(a) 4 gallons to litres.

18 litres

(b) 350 litres to gallons.

35 litres = 7.6 gallons so
350 litres = 76 gallons

2 This graph shows the relationship between speed and time of travel.

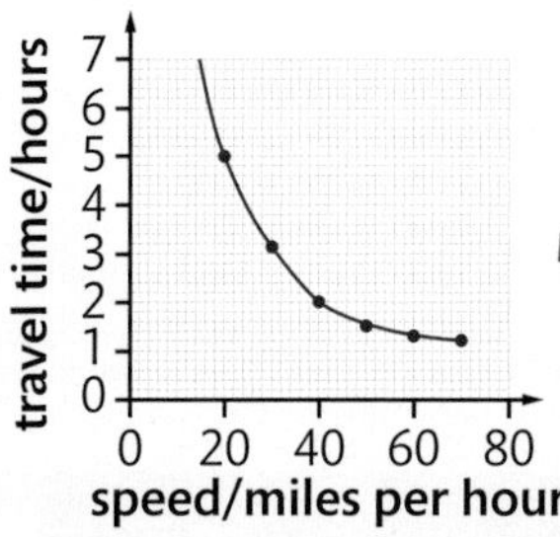

(a) A journey took 5 hours. Use the graph to work out the speed, in miles per hour.

20 miles per hour

(b) The average speed of a journey was 60 miles per hour. Use the graph to find the time taken, in hours.

1.3 hours

Draw lines on the graph to show the values you are working out. To convert 4 gallons to litres draw across from 4 on the vertical axis to the line, then down to the horizontal axis.

Always take readings from graphs accurate to the nearest small square.

Problem solving

The horizontal axis does not go up to 350 litres. But the graph is a **straight line through the origin** so you know that the quantities are directly proportional. This means that the quantities are in a constant ratio. Use the graph to convert 35 litres to gallons then multiply both quantities by 10.

35 : 7.6
× 10   × 10
350 : 76

This is an **inverse proportion** graph. When one quantity increases, the other decreases.

2 Checklist

- A graph of direct proportion goes through the origin.
- Direct proportion has equation $y = kx$.
- A graph of inverse proportion is a curve with asymptotes at the *x*- and *y*-axes.
- Inverse proportion has equation $y = \frac{k}{x}$.

10 Exam-style practice — Grade 6

1 *A* and *B* are directly proportional. Which equation best describes the relationship between *A* and *B*?

$A = 3B + 2 \quad A = \frac{4}{x} \quad A = 6B$

Give a reason for your answer. **[2 marks]**

2 The resistance, *R* ohms, of a wire is inversely proportional to its cross-sectional area, $A\text{ cm}^2$. Given that $R = 60$ when $A = 0.2$, work out the value of *R* when $A = 0.3$. **[2 marks]**

Proportionality formulae

Direct proportion and inverse proportion (see page 53) can be shown algebraically. The symbol $\propto$ means 'is proportional to'.

 ⑤ Proportionality in algebra

Direct proportion

You can write 'y is directly proportional to x' as

$$y \propto x$$

You can replace the proportionality sign with a constant, such as k, to form an equation.

$$y = kx$$

The constant (k in this example) is the **constant of proportionality**.

Inverse proportion

You can write 'y is inversely proportional to x' as

$$y \propto \frac{1}{x}$$

You can replace the proportionality sign with a constant, such as k, to form an equation.

$$y = \frac{k}{x}$$

Again, the constant (k in this example) is the **constant of proportionality**.

⑤ Worked example **Grade 7** ✓

The weight, w, of a piece of wire is directly proportional to its length, l. A piece of wire is 45 cm long and has a weight of 9 grams.

(a) Derive a formula for w in terms of l.

$w \propto l$ so $w = kl$

$9 = k \times 45$ so $k = \dfrac{9}{45} = \dfrac{1}{5}$

$w = \dfrac{1}{5}l$

(b) Work out the value of w when $l = 7.8$ cm.

$w = \dfrac{1}{5}l$

$= \dfrac{1}{5} \times 7.8$

$= 1.56$ grams

(c) Work out the value of l when $w = 22$ g.

$w = \dfrac{1}{5}l$

$22 = \dfrac{1}{5} \times l$ so $l = 22 \times 5$

$l = 110$ cm

> Use your algebraic skills to work through the steps.

⑤ Worked example **Grade 7** ✓

Some friends share a pack of chocolate beans. When the number of friends $(p) = 15$ the number of chocolate beans each one gets $(b) = 8$.

Work out the value of p when $b = 24$.

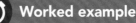

$p \propto \dfrac{1}{b}$ so $p = \dfrac{k}{b}$

$15 = \dfrac{k}{8}$

$k = 15 \times 8 = 120$ so $p = \dfrac{120}{b}$

$p = \dfrac{120}{b}$

$= \dfrac{120}{24}$

$= 5$

> Use the information you are given to work out the value of k, then write the formula.

> Use the formula to work out the required value of p.

> The more friends there are, the fewer chocolate beans each will receive, so p is inversely proportional to b.

 ⑮ Exam-style practice **Grade 7** ✓

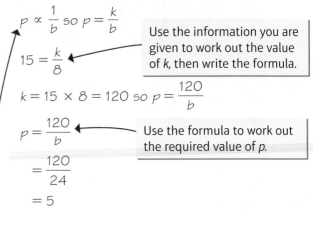

❶ In a spring, the tension (T newtons) is directly proportional to its extension (x cm).

When the tension is 125 newtons, the extension is 5 cm.

(a) Work out a formula for T in terms of x. **[3 marks]**

(b) Work out the tension, in newtons, when the extension is 18 cm. **[1 mark]**

❷ f is inversely proportional to d. When $d = 50$, $f = 225$. What is the value of f when $d = 70$? **[4 marks]**

❸ D is directly proportional to x.

x	2	a
D	32	5

Work out the value of a when $D = 5$. **[4 marks]**

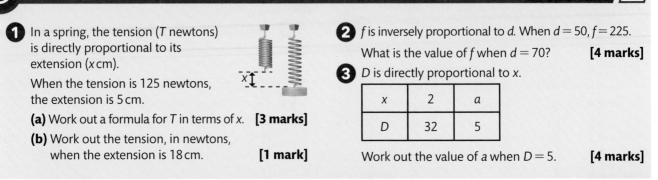

Made a start   Feeling confident   Exam ready

Harder relationships

Direct proportion and inverse proportion may involve square roots, square and cubic relationships.

5 Proportion with squares, square roots and cubics

Direct proportion

You can write 'y is directly proportional to the square of x' as $y \propto x^2$.

You can write 'y is directly proportional to the square root of x' as $y \propto \sqrt{x}$.

You can also write these as $y = kx^2$ and $y = k\sqrt{x}$.

Inverse proportion

$y \propto \frac{1}{x^2}$ means 'y is inversely proportional to the square of x'.

$y \propto \frac{1}{x^3}$ means 'y is inversely proportional to the cube of x'.

You can also write these as $y = \frac{k}{x^2}$ and $y = \frac{k}{x^3}$.

5 Worked example — Grade 8

When a photograph is taken, the exposure time, t seconds, is directly proportional to the square of f.

$t = 0.04$ when $f = 10$.

(a) Derive a formula for t in terms of f.

$t \propto f^2$

$t = kf^2$

$0.04 = k(10)^2$

$0.04 = 100k$

$k = 0.0004$

$t = 0.0004f^2$

Replace the proportionality sign with k, then use the initial values to work out k.

(b) Work out the value of f when $t = 0.0025$.

$t = 0.0004f^2$

$0.0025 = 0.0004f^2$

$f^2 = \frac{0.0025}{0.0004}$

$= 6.25$

$f = 2.5$

Substitute into the formula you have derived.
Use your algebraic skills to work through the steps.

5 Worked example — Grade 8

The pressure, P pascals, of water leaving a hosepipe, is inversely proportional to the square of the radius, R cm, of the hosepipe.

$P = 30$ when $R = 3$.

(a) Derive a formula for P in terms of R.

$P \propto \frac{1}{R^2}$

$P = \frac{k}{R^2}$

$30 = \frac{k}{3^2}$

$k = 270$

$P = \frac{270}{R^2}$

Replace the proportionality sign with k, then use the initial values to work out k.

(b) Work out the value of P when $R = 2.5$.

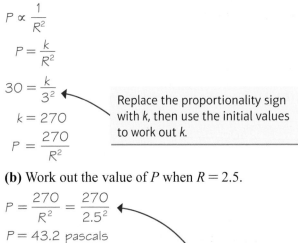

$P = \frac{270}{R^2} = \frac{270}{2.5^2}$

$P = 43.2$ pascals

Substitute into the formula you have derived.

10 Exam-style practice — Grade 8

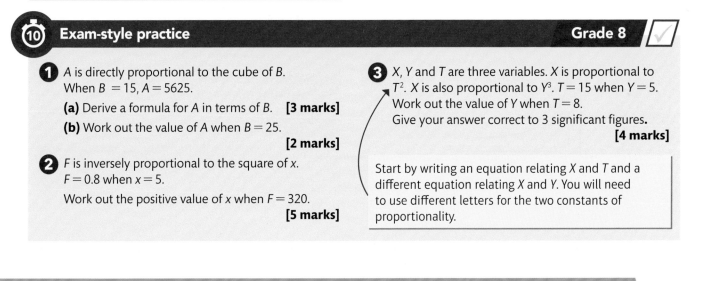

1 A is directly proportional to the cube of B.
When $B = 15$, $A = 5625$.

(a) Derive a formula for A in terms of B. **[3 marks]**

(b) Work out the value of A when $B = 25$. **[2 marks]**

2 F is inversely proportional to the square of x.
$F = 0.8$ when $x = 5$.

Work out the positive value of x when $F = 320$. **[5 marks]**

3 X, Y and T are three variables. X is proportional to T^2. X is also proportional to Y^3. $T = 15$ when $Y = 5$.
Work out the value of Y when $T = 8$.
Give your answer correct to 3 significant figures. **[4 marks]**

Start by writing an equation relating X and T and a different equation relating X and Y. You will need to use different letters for the two constants of proportionality.

Ratio and proportion

Read the exam-style question and worked solution, then practise your exam skills with the two questions at the bottom of the page.

⑩ Worked example — Grade 7 ✓

The magnetic force, F newtons, is inversely proportional to the square of the distance, r cm, between two magnets.

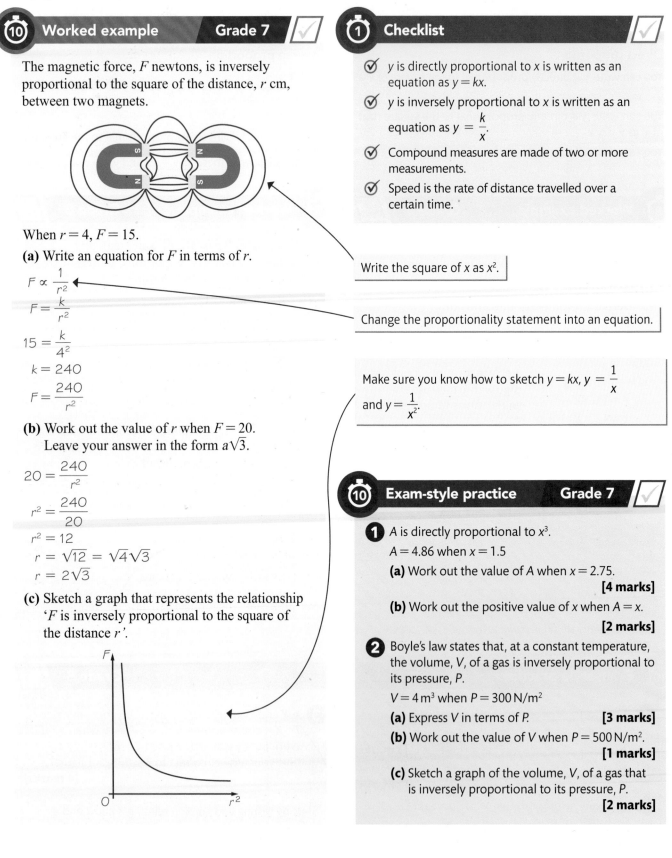

When $r = 4$, $F = 15$.

(a) Write an equation for F in terms of r.

$$F \propto \frac{1}{r^2}$$
$$F = \frac{k}{r^2}$$
$$15 = \frac{k}{4^2}$$
$$k = 240$$
$$F = \frac{240}{r^2}$$

(b) Work out the value of r when $F = 20$.
Leave your answer in the form $a\sqrt{3}$.

$$20 = \frac{240}{r^2}$$
$$r^2 = \frac{240}{20}$$
$$r^2 = 12$$
$$r = \sqrt{12} = \sqrt{4}\sqrt{3}$$
$$r = 2\sqrt{3}$$

(c) Sketch a graph that represents the relationship 'F is inversely proportional to the square of the distance r'.

① Checklist ✓

☑ y is directly proportional to x is written as an equation as $y = kx$.

☑ y is inversely proportional to x is written as an equation as $y = \dfrac{k}{x}$.

☑ Compound measures are made of two or more measurements.

☑ Speed is the rate of distance travelled over a certain time.

> Write the square of x as x^2.

> Change the proportionality statement into an equation.

> Make sure you know how to sketch $y = kx$, $y = \dfrac{1}{x}$ and $y = \dfrac{1}{x^2}$.

⑩ Exam-style practice — Grade 7 ✓

1 A is directly proportional to x^3.
$A = 4.86$ when $x = 1.5$
(a) Work out the value of A when $x = 2.75$.
[4 marks]
(b) Work out the positive value of x when $A = x$.
[2 marks]

2 Boyle's law states that, at a constant temperature, the volume, V, of a gas is inversely proportional to its pressure, P.
$V = 4\,m^3$ when $P = 300\,N/m^2$
(a) Express V in terms of P. **[3 marks]**
(b) Work out the value of V when $P = 500\,N/m^2$.
[1 marks]
(c) Sketch a graph of the volume, V, of a gas that is inversely proportional to its pressure, P.
[2 marks]

Made a start ✓   Feeling confident ✓   Exam ready

Angle properties

You need to be able to remember and use angle properties involving intersecting lines and parallel lines.

5 Properties of angles

Alternate angles are equal.

$a = b$

Corresponding angles are equal.

$a = b$

Allied or **co-interior** angles sum to 180°.

$a + b = 180°$

Angles on a straight line sum to 180°.

$a + b + c = 180°$

Vertically opposite angles are equal.

$a = c$ and $b = d$

Angles around a point add up to 360°.

$a + b + c + d = 360°$

10 Worked example   Grade 6

Work out the size of each unknown angle in this parallelogram. Give a reason for each answer.

(a) a

$a = 180° - 110° = 70°$ as angles on a straight line add up to 180°.

(b) b

$b = 70°$ as a and b are alternate angles so they are equal.

(c) c

$c = 110°$ as angles 110° and c are corresponding angles and are equal.

(d) d

$d = 180° - 70° = 110°$ as b and d are allied angles so add up to 180°.

(e) e

$e = 110°$ as angles 110° and e are alternate angles so they are equal.

3 Proof

The base of the triangle is parallel to the line, so angle A = angle X and angle C = angle Y.

Angles $X + B + Y = 180°$, since angles on a straight line add up to 180°.

Hence, angles in the triangle, $A + B + C = 180°$.

Problem solving

Always start by looking for angles that are related to known angles, then use these to find further unknown angles.

You could have worked out the angles in a different way, for example 110° and b are allied angles that add up to 180°.

15 Exam-style practice   Grade 6

1 *AFB* and *CHD* are parallel lines. *EFD* is a straight line. Work out the size of angle x.

[3 marks]

2 *ABC* and *DEF* are parallel lines. *BEG* is a straight line. Angle *GEF* = 47°. Work out the value of x and give reasons.

[3 marks]

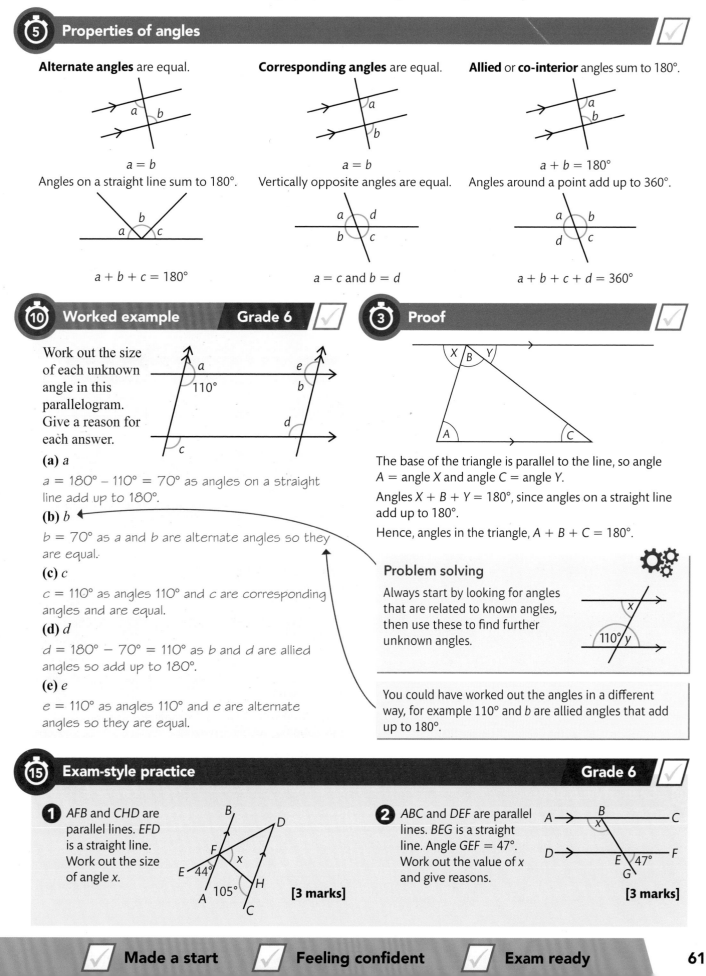

Angle problems

When you solve angle problems you must give a valid reason for each step of your working.

② Angles checklist

You need to learn these rules.

- ✓ Corresponding angles are equal.
- ✓ Alternate angles are equal.
- ✓ Allied or co-interior angles add up to 180°.
- ✓ Angles in a triangle add up to 180°.
- ✓ Base angles in an isosceles triangle are equal.
- ✓ Opposite angles are equal.
- ✓ Angles around a point add up to 360°.
- ✓ Angles on a straight line add up to 180°.

You often need to find other angles before you find the required angle.

Start by listing the angles that you know are correct and write a reason in each case. Angle ABE is a good starting point.

⑤ Worked example — Grade 6

ABC, DEF and $PQRS$ are parallel lines.

BEQ is a straight line.

Angle $ABE = 65°$ and angle $QER = 85°$.

Work out the size of angle x.

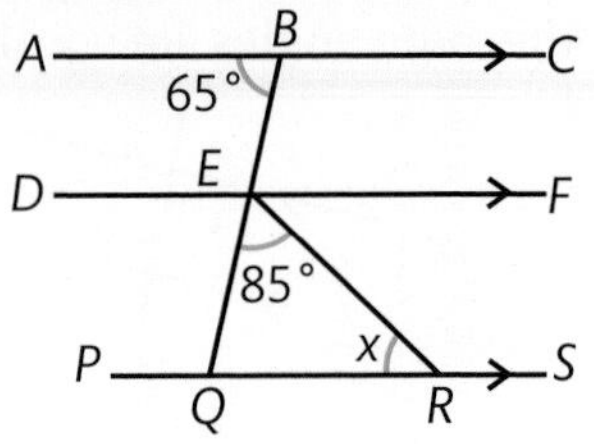

Give reasons for each stage of your working.

Angle DEQ = angle $ABE = 65°$
(corresponding angles are equal)
Angle $FER = 180° - (65° + 85°) = 30°$
(angles on a straight line add up to 180°)
x = angle $FER = 30°$
(alternate angles are equal)

⑩ Worked example — Grade 6

❶ ABC is parallel to DEF. EBP is a straight line.

$AB = EB$. Angle $PBC = 48°$ and angle $AED = y°$.

Work out the value of y.

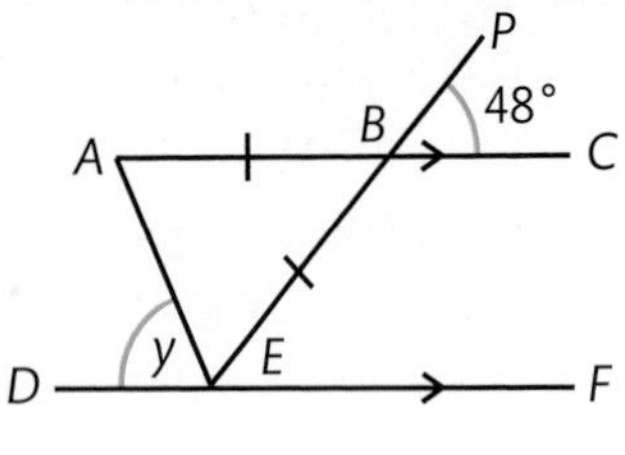

Give a reason for each stage of your working.

Angle ABE = angle $PBC = 48°$
(opposite angles are equal)
Angle BEF = angle $ABE = 48°$
(alternate angles are equal)
Angle BAE + angle $BEA = 180° - 48° = 132°$
(angles in a triangle add up to 180°)
Angle $BEA = 132° ÷ 2 = 66°$
(base angles in an isosceles triangle are equal)
$y = 180° - (66° + 48°) = 66°$
(angles on a straight line add up to 180°)

❷ BDE is an isosceles triangle. $DB = DE$.

The straight line ABC is parallel to the straight line DEF.

Work out the size of the angle BDE in terms of x.

You must give reasons for each stage in your working.

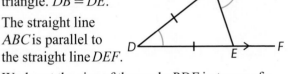

Angle CBE = angle $BED = x$
(alternate angles are equal)
Angle $DBE = x$
(base angles in an isosceles triangle are equal)
Angle $BDE = 180° - 2x$
(angles in a triangle sum to 180°)

Exam focus 📌

Write down every step of your working, whether it is a calculation or notes to yourself.

⑩ Exam-style practice — Grade 6

$CDEF$ is a straight line. AB is parallel to CF. $DE = AE$.

Work out the size of angle x.

You must give reasons for your answer.

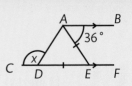

[4 marks]

☑ Made a start   ☑ Feeling confident   ☑ Exam ready

Angles in polygons

A polygon is a closed two-dimensional shape with three or more straight sides, such as a triangle, quadrilateral or hexagon. You can work out facts about the **interior angles** and **exterior angles** of a polygon. For instance, the sum of the interior angles of a quadrilateral will always be 360°.

5 Regular polygons

All the sides of a regular polygon are equal and all its angles are equal. The number of sides in a polygon is represented by n.

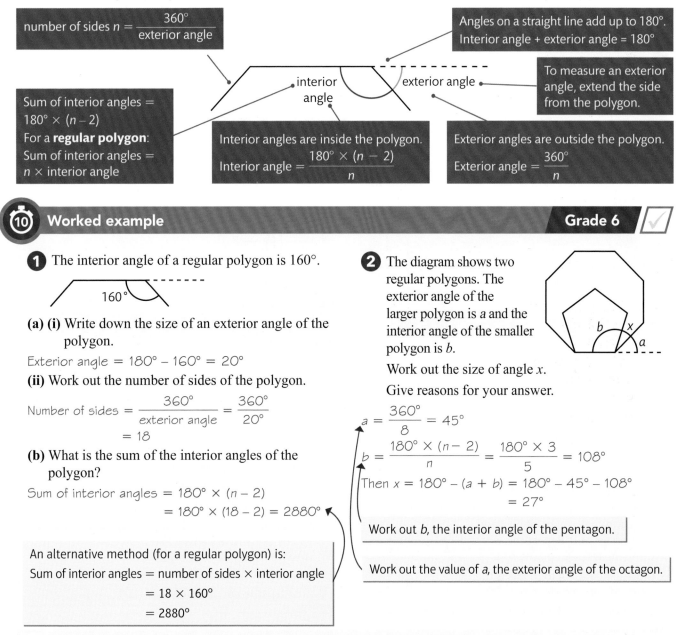

number of sides $n = \dfrac{360°}{\text{exterior angle}}$

Angles on a straight line add up to 180°.
Interior angle + exterior angle = 180°

To measure an exterior angle, extend the side from the polygon.

Sum of interior angles = 180° × (n – 2)
For a **regular polygon**:
Sum of interior angles = n × interior angle

Interior angles are inside the polygon.
Interior angle = $\dfrac{180° \times (n - 2)}{n}$

Exterior angles are outside the polygon.
Exterior angle = $\dfrac{360°}{n}$

10 Worked example — Grade 6

1 The interior angle of a regular polygon is 160°.

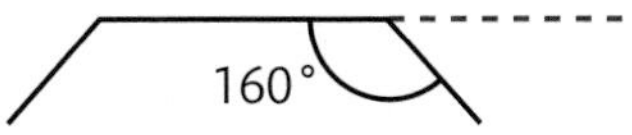

(a) (i) Write down the size of an exterior angle of the polygon.

Exterior angle = 180° – 160° = 20°

(ii) Work out the number of sides of the polygon.

Number of sides = $\dfrac{360°}{\text{exterior angle}} = \dfrac{360°}{20°}$
= 18

(b) What is the sum of the interior angles of the polygon?

Sum of interior angles = 180° × (n – 2)
= 180° × (18 – 2) = 2880°

An alternative method (for a regular polygon) is:
Sum of interior angles = number of sides × interior angle
= 18 × 160°
= 2880°

2 The diagram shows two regular polygons. The exterior angle of the larger polygon is a and the interior angle of the smaller polygon is b.

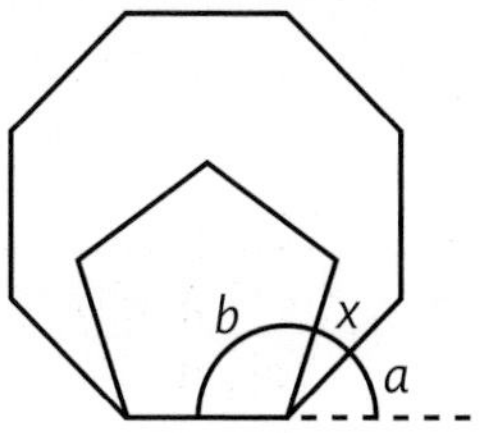

Work out the size of angle x.
Give reasons for your answer.

$a = \dfrac{360°}{8} = 45°$

$b = \dfrac{180° \times (n - 2)}{n} = \dfrac{180° \times 3}{5} = 108°$

Then x = 180° – (a + b) = 180° – 45° – 108°
= 27°

Work out b, the interior angle of the pentagon.

Work out the value of a, the exterior angle of the octagon.

15 Exam-style practice — Grades 5–6

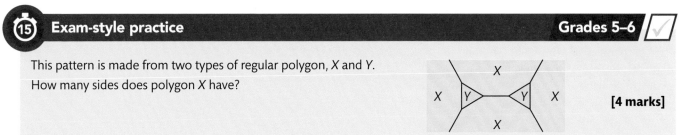

This pattern is made from two types of regular polygon, X and Y.
How many sides does polygon X have?

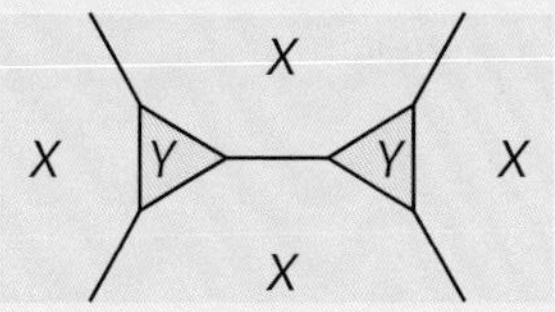

[4 marks]

Constructing perpendiculars

You must be able to construct perpendicular lines using a pencil, a pair of compasses and a ruler. Remember that the perpendicular distance from a point to a line is the shortest distance between them.

5 Constructing perpendicular lines

You need to know three constructions involving perpendicular lines:

1. the midpoint and **perpendicular bisector** of a line segment
2. the perpendicular from a point to a line
3. the perpendicular from a point on a line.

To **bisect** means to cut something in half.

1. Place the compass point at *A* and then draw an arc.
2. Place the compass point at *B* and then draw an arc.
3. Join the two points where the arcs meet.

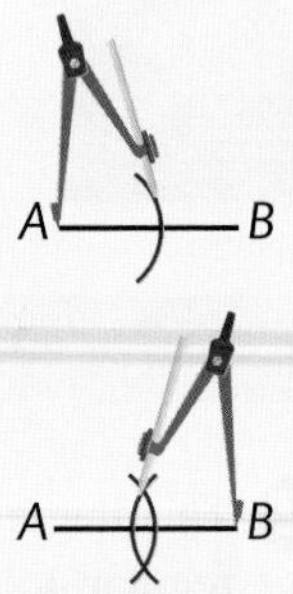

2 Checklist

- Make sure the joint of your pair of compasses is not loose and your pencil is clamped firmly in place.
- Use a transparent ruler and a sharp pencil.
- Always label your angles and sides.
- Never rub out your construction lines.

5 Worked example Grade 5

Use a ruler and compasses to construct the perpendicular bisector of the line segment *AB*. You must show all your construction lines.

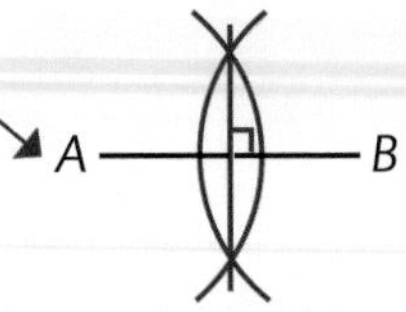

5 Worked example Grade 5

Use a ruler and compasses to construct the perpendicular from point *A* to the line *XY*. You must show all your construction lines.

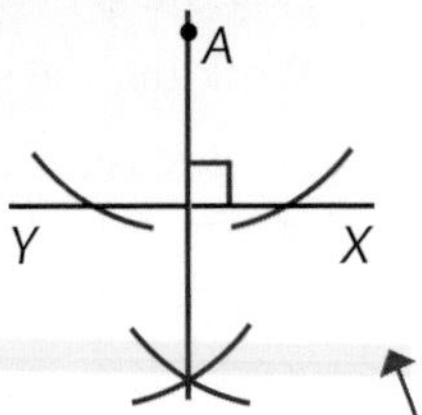

1. Draw two arcs from *A* to cut the line at *B* and *C*.
2. Place the compass point on *B* and draw an arc at *D*. Repeat this with the compass point placed on *C*.
3. Draw a line from *A* to *D*. This is the required perpendicular line.

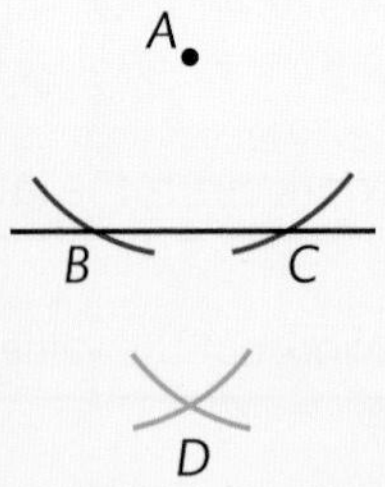

5 Worked example Grade 5

Use a ruler and compasses to construct the perpendicular at point *P* on the line *AB*. You must show all your construction lines.

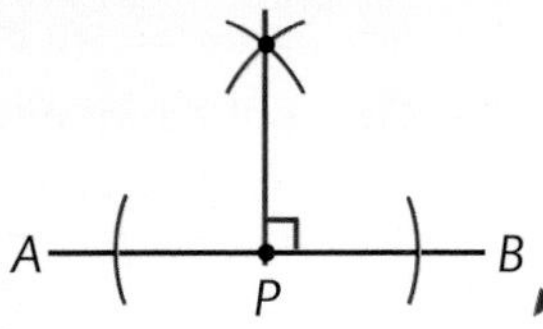

1. Place the compass at *P* and then draw arcs of equal length from *P* to cross *AB* at *C* and *D*
2. Extend the compass and place its point on *C*. Draw an arc and repeat this from *D*. These arcs intersect at *E*.
3. Draw a perpendicular line from *E* to *P*.

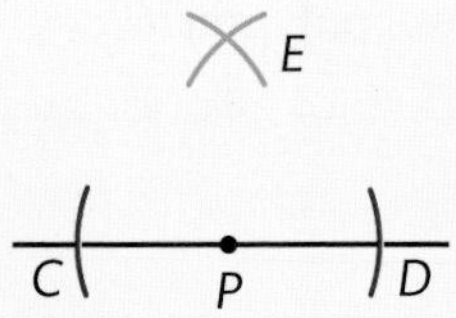

5 Exam-style practice Grade 5

Use a ruler and a compass to **construct** the perpendicular from point *C* to the line *AB*. You must show all your construction lines.

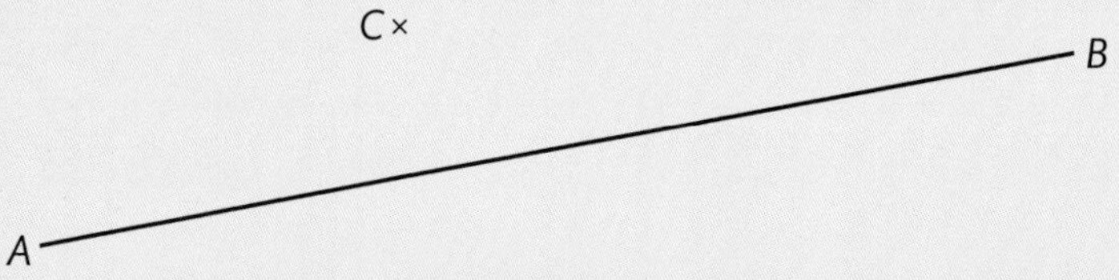

[2 marks]

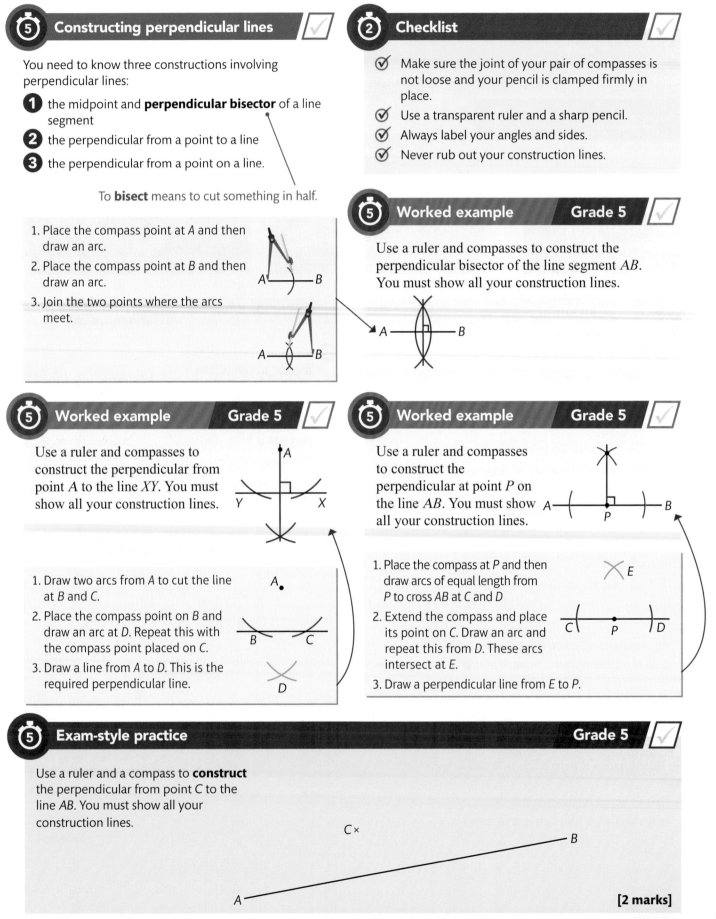

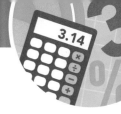

Constructions, plans and elevations

You need to be able to draw or interpret a 2-D drawing of a 3-D object from different angles. These are called plans and elevations. You also need to know how to bisect angles.

5 Bisecting an angle

You can use a pencil, a pair of compasses and a ruler to bisect an angle by following three simple steps:

1. Set the compasses to a radius of a length shorter than the lines of the angle. With the compass point on *A*, sweep an arc and mark points *B* and *C*.

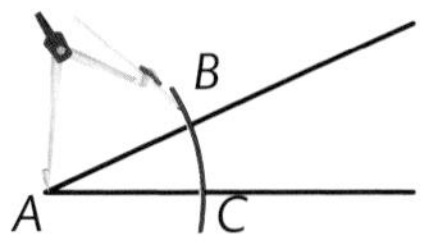
B
A
C

2. Using the same radius, draw arcs from points *B* and *C* to intersect at *D*.

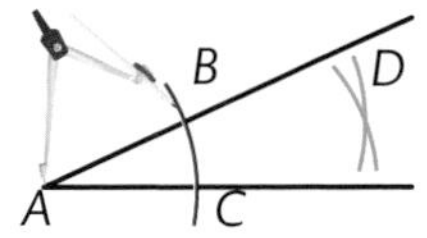
B
D
A
C

3. Draw a line from *A* to *D* to bisect the angle *CAB*.

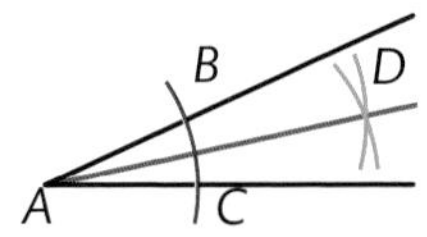
B
D
A
C

5 Worked example — Grade 5

Here is an equilateral triangular prism.

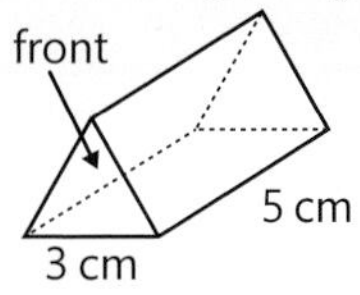
front
5 cm
3 cm

Construct the plan and front elevation of the equilateral triangular prism.

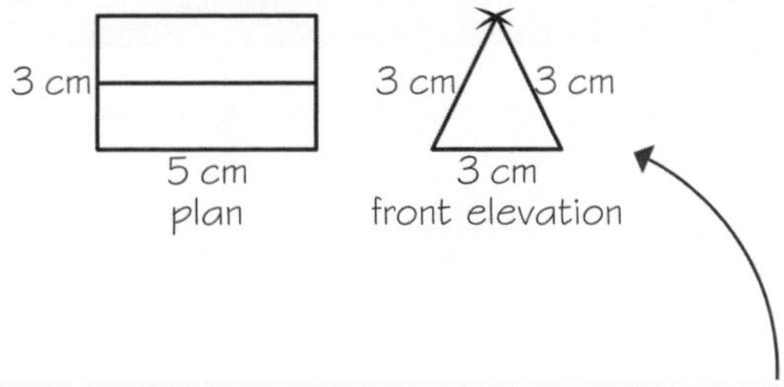
3 cm
5 cm
plan
3 cm
3 cm
3 cm
front elevation

The front elevation is an equilateral triangle. Draw a 3 cm line using a ruler and pencil. Then set your compasses to 3 cm and draw arcs with the compass point at each end of your line. The point where these arcs cross is the third vertex of the triangle.

2 Constructing 45° and 30° angles

A right angle can be bisected to construct a 45° angle and, in a similar way, a 60° angle can be bisected to construct a 30° angle.

5 Plans and elevations

The diagram below shows the different sides of a three-dimensional shape. Each view has a special name.

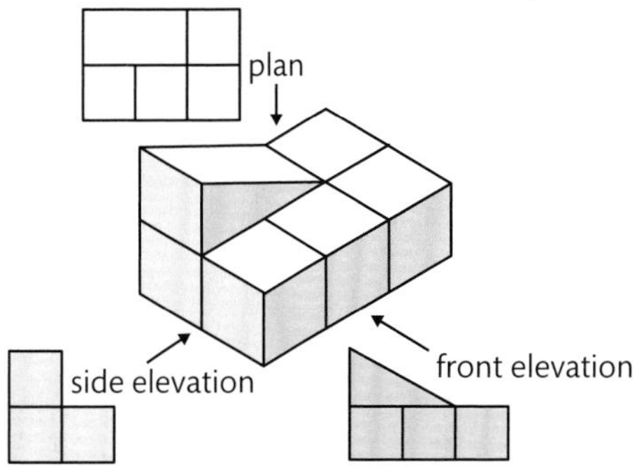
plan
side elevation
front elevation

The **plan view** is the view looking down onto the shape from above. The **front elevation** is the view looking at the shape from the front. The **side elevation** is the view looking at the shape from the side.

15 Exam-style practice — Grade 6

1. Construct an equilateral triangle with sides of length 6 cm. You must show all your construction lines. **[2 marks]**
2. **(a)** Draw a straight line and construct its perpendicular bisector. **[2 marks]**
   (b) By bisecting one of the right angles on your diagram, construct and label an angle of 45°. **[2 marks]**
3. The plan, front elevation and side elevation of a solid prism are drawn on a centimetre grid.

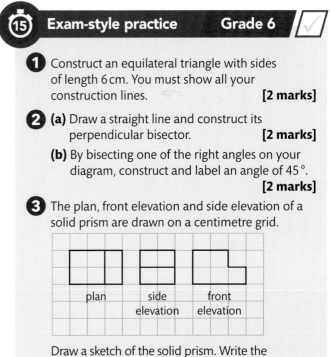

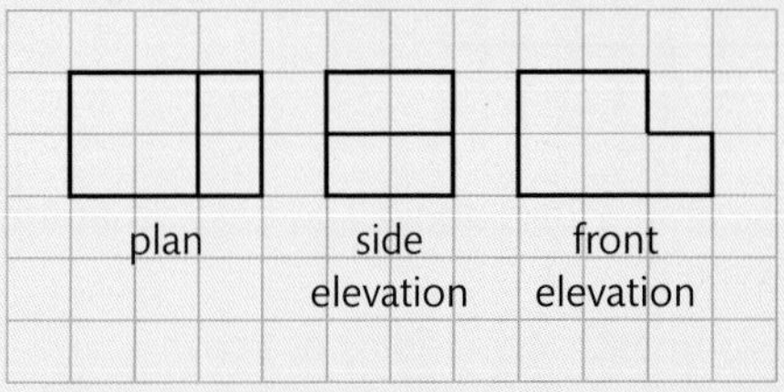
plan
side elevation
front elevation

Draw a sketch of the solid prism. Write the dimensions of the prism on your sketch. **[2 marks]**

Loci

A **locus** is a line or a path. The line or path is formed by a point that moves according to a particular rule. You will need to apply these rules with some of the techniques for constructing lines. The plural of locus is loci.

⑤ Common loci ✓

The locus of points that are a constant distance from one point is a circle.

The locus of points that are equidistant from two fixed points is the perpendicular bisector of the line joining the points.

The locus of points that are at a constant distance from a fixed line consists of two straight lines parallel to the original line, and two semicircles.

The locus of a point that is equidistant from two intersecting lines is the angle bisector of the lines.

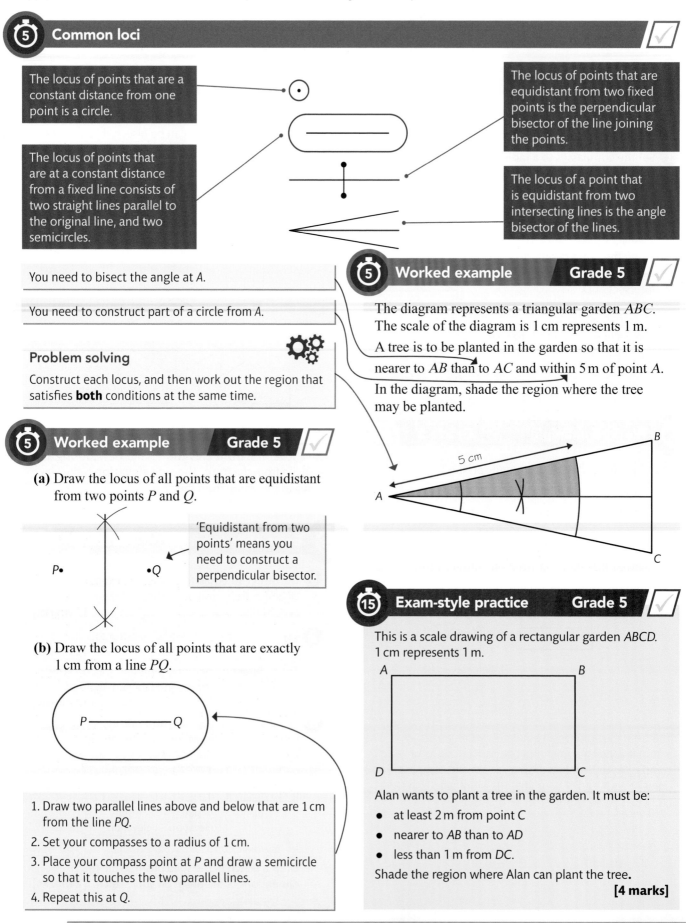

You need to bisect the angle at *A*.

You need to construct part of a circle from *A*.

Problem solving

Construct each locus, and then work out the region that satisfies **both** conditions at the same time.

⑤ Worked example Grade 5 ✓

The diagram represents a triangular garden *ABC*. The scale of the diagram is 1 cm represents 1 m.

A tree is to be planted in the garden so that it is nearer to *AB* than to *AC* and within 5 m of point *A*. In the diagram, shade the region where the tree may be planted.

5 cm

⑤ Worked example Grade 5 ✓

(a) Draw the locus of all points that are equidistant from two points *P* and *Q*.

'Equidistant from two points' means you need to construct a perpendicular bisector.

P•   *•Q*

(b) Draw the locus of all points that are exactly 1 cm from a line *PQ*.

P————*Q*

1. Draw two parallel lines above and below that are 1 cm from the line *PQ*.
2. Set your compasses to a radius of 1 cm.
3. Place your compass point at *P* and draw a semicircle so that it touches the two parallel lines.
4. Repeat this at *Q*.

⑮ Exam-style practice Grade 5 ✓

This is a scale drawing of a rectangular garden *ABCD*. 1 cm represents 1 m.

A   *B*

D   *C*

Alan wants to plant a tree in the garden. It must be:
- at least 2 m from point *C*
- nearer to *AB* than to *AD*
- less than 1 m from *DC*.

Shade the region where Alan can plant the tree.

[4 marks]

✓ **Made a start**   ✓ **Feeling confident**   ✓ **Exam ready**

Perimeter and area

Problems about perimeter and area can be solved by algebraic methods.

 (2) Perimeter and area ✓

The **perimeter** of a shape is the total distance around its edges. It is usually measured in mm, cm, m or km.

The **area** of a shape is a measure of the two-dimensional space that it covers. It is usually measured in mm², cm², m² or km².

---

(5) Triangles and trapezia xy^2 ✓

You will need to remember the formulae for the areas of triangles, trapeziums and parallelograms.

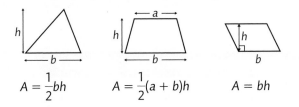

$$A = \frac{1}{2}bh \qquad A = \frac{1}{2}(a + b)h \qquad A = bh$$

---

 (5) Worked example **Grades 6–7** ✓

The diagram shows a trapezium.

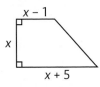

All the measurements are in centimetres.

The area of the trapezium is 63 cm².

(a) Show that $x^2 + 2x - 63 = 0$.

Area of trapezium $= \dfrac{1}{2}(a + b)h$

$$= \frac{1}{2}(x - 1 + x + 5)x$$
$$= \frac{1}{2}(2x + 4)x$$
$$= x^2 + 2x$$
$$x^2 + 2x = 63$$
$$x^2 + 2x - 63 = 0$$

(b) Work out the value of x.

$$x^2 + 2x - 63 = 0$$
$$(x + 9)(x - 7) = 0$$
$$x = -9 \text{ or } x = 7$$

The value of x is 7 as x cannot have a negative value.

---

(5) Worked example **Grades 6–7** ✓

All the measurements in the diagram are in centimetres. The area of the shape is A cm².

Work out a formula for A in terms of x. You must write your formula as simply as possible.

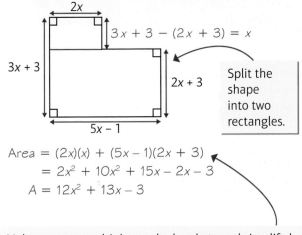

$3x + 3 - (2x + 3) = x$

Split the shape into two rectangles.

Area $= (2x)(x) + (5x - 1)(2x + 3)$
$= 2x^2 + 10x^2 + 15x - 2x - 3$
$A = 12x^2 + 13x - 3$

Make sure you multiply out the brackets and simplify by collecting like terms.

---

(10) Exam-style practice **Grade 6** ✓

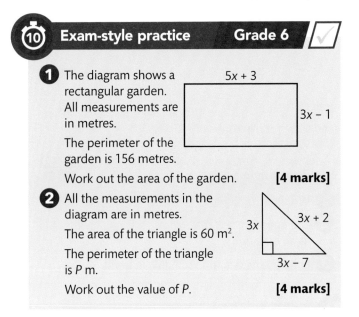

1 The diagram shows a rectangular garden. All measurements are in metres.

5x + 3
3x - 1

The perimeter of the garden is 156 metres.

Work out the area of the garden. **[4 marks]**

2 All the measurements in the diagram are in metres.

The area of the triangle is 60 m².

3x
3x + 2
3x - 7

The perimeter of the triangle is P m.

Work out the value of P. **[4 marks]**

Volumes of 3D shapes

The volume and surface area of a 3D shape can be calculated from simple formulae. You will be given these in the exam but you need to recognise when you need them and be able to manipulate them.

5 Cones, spheres and pyramids xy^2

If you need to use one of these formulae in your exam, it will be given with the question.

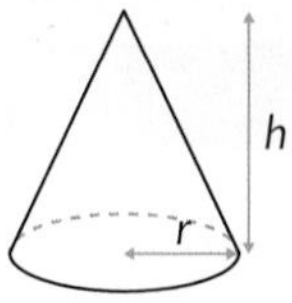

$V = \frac{1}{3}\pi r^2 h$

$V = \frac{1}{3} \times$ area of base $\times$ vertical height

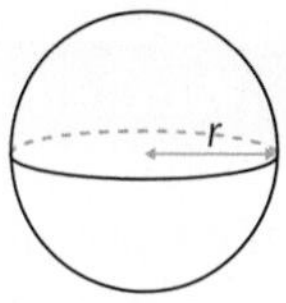

$V = \frac{4}{3}\pi r^3$

$V = \frac{4}{3} \times \pi \times (\text{radius})^3$

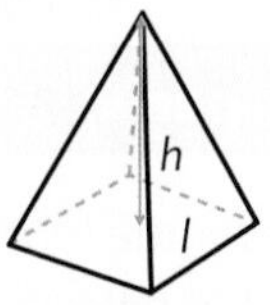

The length of the base is squared here because the base is a square.

$V = \frac{1}{3}l^2 h$

$V = \frac{1}{3} \times$ area of base $\times$ vertical height

5 Worked example — Grade 7

The diagram shows a cylinder and a sphere.

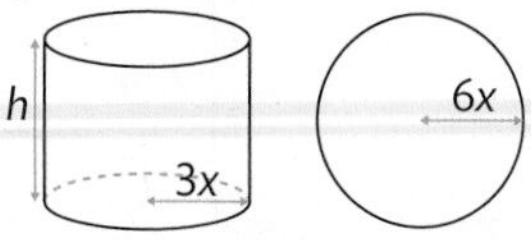

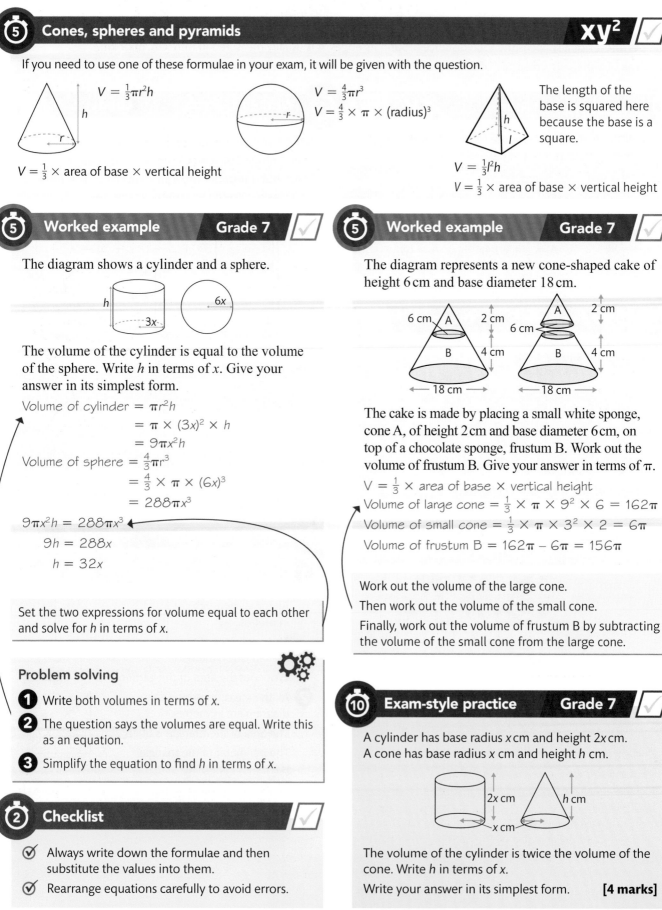

The volume of the cylinder is equal to the volume of the sphere. Write h in terms of x. Give your answer in its simplest form.

Volume of cylinder $= \pi r^2 h$

$= \pi \times (3x)^2 \times h$

$= 9\pi x^2 h$

Volume of sphere $= \frac{4}{3}\pi r^3$

$= \frac{4}{3} \times \pi \times (6x)^3$

$= 288\pi x^3$

$9\pi x^2 h = 288\pi x^3$

$9h = 288x$

$h = 32x$

Set the two expressions for volume equal to each other and solve for h in terms of x.

Problem solving

1. Write both volumes in terms of x.
2. The question says the volumes are equal. Write this as an equation.
3. Simplify the equation to find h in terms of x.

2 Checklist

- Always write down the formulae and then substitute the values into them.
- Rearrange equations carefully to avoid errors.

5 Worked example — Grade 7

The diagram represents a new cone-shaped cake of height 6 cm and base diameter 18 cm.

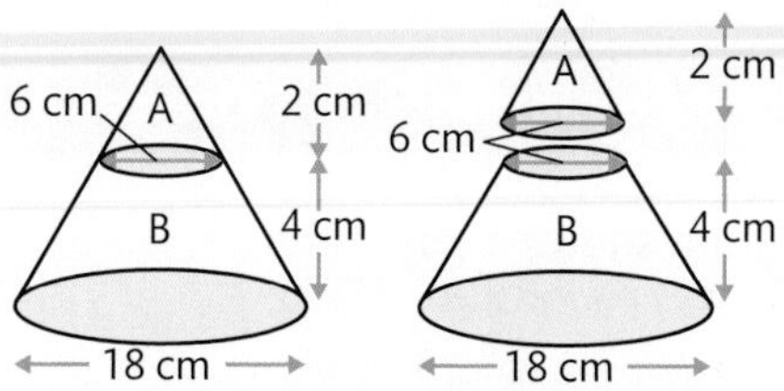

The cake is made by placing a small white sponge, cone A, of height 2 cm and base diameter 6 cm, on top of a chocolate sponge, frustum B. Work out the volume of frustum B. Give your answer in terms of π.

$V = \frac{1}{3} \times$ area of base $\times$ vertical height

Volume of large cone $= \frac{1}{3} \times \pi \times 9^2 \times 6 = 162\pi$

Volume of small cone $= \frac{1}{3} \times \pi \times 3^2 \times 2 = 6\pi$

Volume of frustum B $= 162\pi - 6\pi = 156\pi$

Work out the volume of the large cone.
Then work out the volume of the small cone.
Finally, work out the volume of frustum B by subtracting the volume of the small cone from the large cone.

10 Exam-style practice — Grade 7

A cylinder has base radius x cm and height $2x$ cm.
A cone has base radius x cm and height h cm.

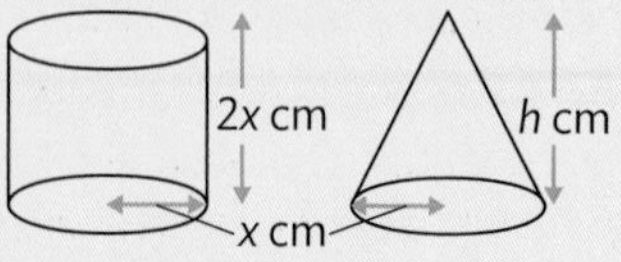

The volume of the cylinder is twice the volume of the cone. Write h in terms of x.

Write your answer in its simplest form. **[4 marks]**

Surface area

Surface area is the total area of all the exterior surfaces of a three-dimensional object.

5 Cones and spheres xy²

The surface area of a cone is made up of the area of the curved surface and the area of the base.

Area of curved surface $= \pi r l$

Area of base $= \pi r^2$

Total surface area $= \pi r^2 + \pi r l$

l
r

These formulae will be given in the exam. You should know how to substitute numbers into them.

Total surface area of a sphere $= 4\pi r^2$

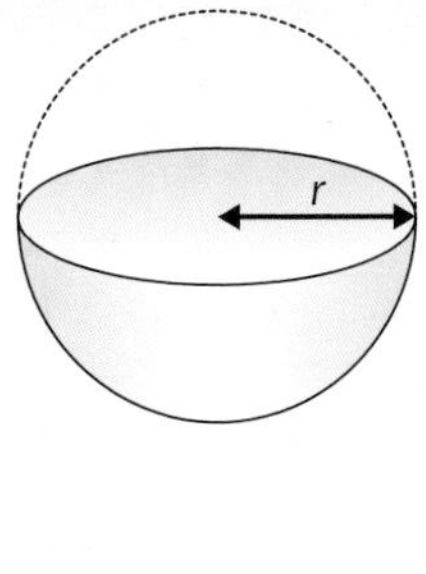

r

A hemisphere is half of a sphere. You need to halve the surface area of a sphere to get the surface area of the curved part of a hemisphere.

10 Worked example Grade 6

1 The diagram represents a cone. The height of the cone is 15 cm. The diameter of the base of the cone is 16 cm. Work out the area of the curved surface of the cone. Give your answer as a multiple of π.

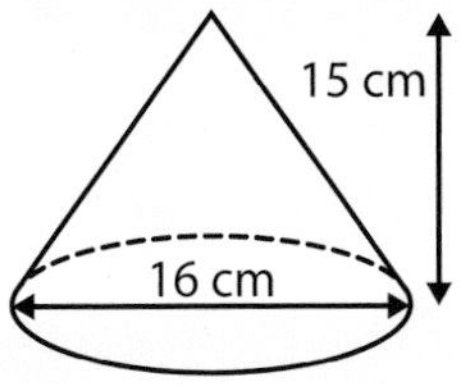

15 cm
16 cm

$l^2 = 8^2 + 15^2$

$l^2 = 64 + 225$

$l^2 = 289$

$l = 17$

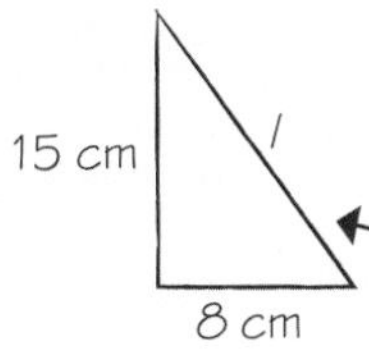

15 cm
l
8 cm

Area of curved surface $= \pi r l$

$= \pi \times 8 \times 17$

$= 136\pi$

You have not been given the slant height or the radius. You will need to work out the slant height by using Pythagoras' theorem.

Do not include the area of the base of the cone as this will be joined to the flat face of the hemisphere and will be an interior face.

2 The diagram shows a solid shape made of a hemisphere of radius 5 cm and a cone of radius 5 cm with a slant height of 13 cm.

Work out the total surface area of the solid shape. Give your answer in terms of π.

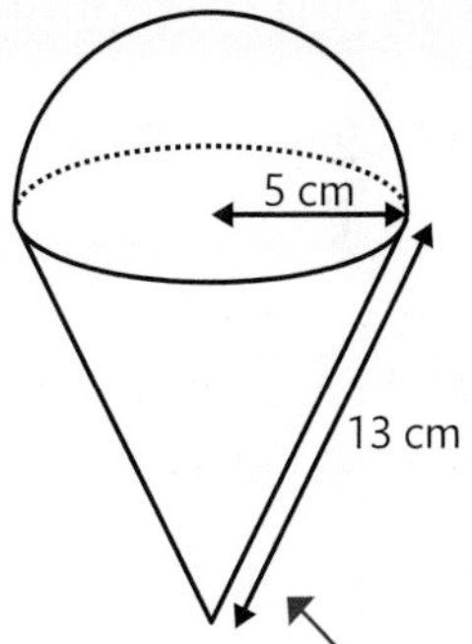

5 cm
13 cm

Area of curved surface of cone

$= \pi r l$

$= \pi \times 5 \times 13$

$= 65\pi$

Surface area of hemisphere $= \frac{1}{2} \times 4\pi r^2$

$= \frac{1}{2} \times 4 \times \pi \times 5^2$

$= \frac{1}{2} \times 4 \times \pi \times 25 = 50\pi$

Total surface area $= 65\pi + 50\pi$

$= 115\pi$

Split the shape into two known shapes and work out the surface area of each shape separately.

2 Checklist

- Compound shapes are made up of two or more shapes.
- To work out the surface area of a 3D compound shape, work out the area of each exterior face or surface and then add them together to find the total surface area.

10 Exam-style practice Grades 5–6

The diagram shows a solid shape made of a hemisphere and cylinder of diameter 8 cm.

Work out the total surface area of the solid shape.

Give your answer in terms of π. **[4 marks]**

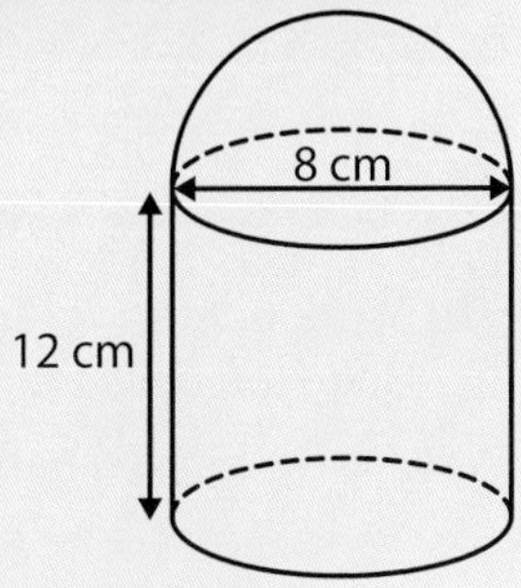

8 cm
12 cm

Prisms

You should be able to work out the surface area and volume of each prism on this page.

5 Facts about prisms

If the cross-section of a prism is a polygon, the other faces will be rectangles. The cross-section of a prism is the same all along its length.

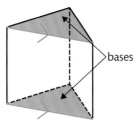

bases

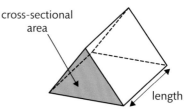

cross-sectional area

length

The bases are parallel to each other.

Adjacent edges of the rectangles are all parallel.

Surface area of prism = sum of areas of all the faces

Volume of prism = cross-sectional area × length

5 Worked example   Grade 5

The diagram shows a triangular prism.

10 cm

8 cm

9 cm

6 cm

(a) Work out the surface area of the triangular prism. State the units with your answer.

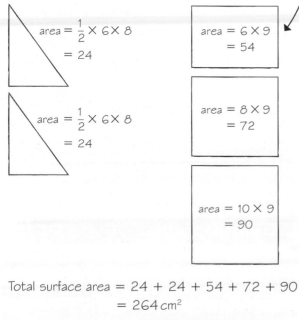

area = $\frac{1}{2}$ × 6 × 8

= 24

area = 6 × 9

= 54

area = $\frac{1}{2}$ × 6 × 8

= 24

area = 8 × 9

= 72

area = 10 × 9

= 90

Total surface area = 24 + 24 + 54 + 72 + 90

= 264 cm²

(b) Work out the volume of the triangular prism. State the units with your answer.

Volume = cross-sectional area × length

= $\frac{1}{2}$ × 6 × 8 × 9 = 216 cm³

Problem solving ⚙

If you need to work out the surface area of a solid, it is a good idea to sketch each face. Count to make sure you haven't missed any faces.

2 Checklist

☑ Sketch each face to work out the total surface area.

☑ Work out the cross-sectional area to find the volume.

10 Exam-style practice   Grade 5

1 The diagram shows a cuboid of dimensions 10 cm × 8 cm × 5 cm.

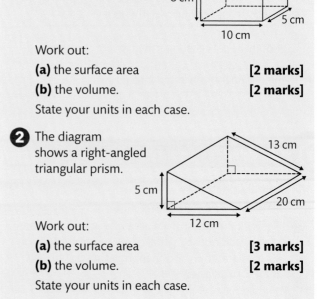

8 cm

5 cm

10 cm

Work out:

(a) the surface area   **[2 marks]**

(b) the volume.   **[2 marks]**

State your units in each case.

2 The diagram shows a right-angled triangular prism.

13 cm

5 cm

20 cm

12 cm

Work out:

(a) the surface area   **[3 marks]**

(b) the volume.   **[2 marks]**

State your units in each case.

Circles and cylinders

A cylinder is a prism with a circular cross-section. The area of the cylinder's circular end is used to calculate the volume.

5 Circles and cylinders xy^2

You need to know these formulae for the exam.

Circle

Area of circle $= \pi r^2$

Circumference of circle $= \pi d = 2\pi r$

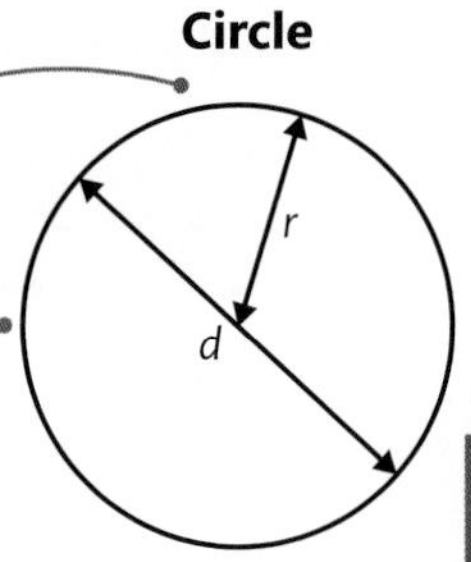

r
d

Cylinder

Area of top $= \pi r^2$

Area of curved surface $= 2\pi rh$

Volume of cylinder $= \pi r^2 h$

Surface area of cylinder $= 2\pi r^2 + 2\pi rh$

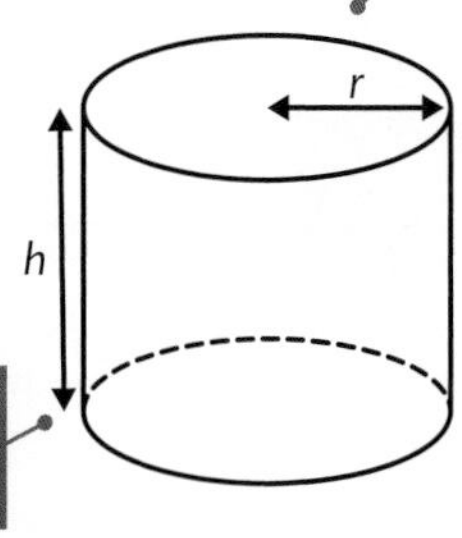

r
h

10 Worked example — Grade 6

1 The diagram shows an open-topped cylinder.

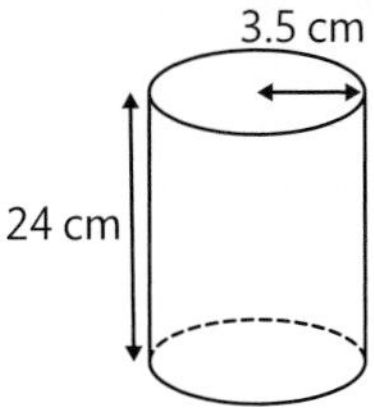

3.5 cm
24 cm

Work out:

(a) the surface area. State the units. Give your answer correct to 3 significant figures.

Surface area $= \pi r^2 + 2\pi rh$
$= (\pi \times 3.5^2) + (2 \times \pi \times 3.5 \times 24)$
$= 566\,\text{cm}^2$

You could also give an exact answer in terms of π.

The surface area is $\frac{721}{4}\pi$.

(b) the volume. State the units. Give your answer correct to 3 significant figures.

Volume of cylinder $= \pi r^2 h$
$= \pi \times 3.5^2 \times 24$
$= 924\,\text{cm}^3$

2 The diameter of Sandeep's bicycle wheel is 0.75 m.

He cycles 500 m.

Work out the number of complete turns the wheel makes.

Circumference $= \pi d = \pi \times 0.75 = 0.75\pi$
Number of turns $= 500 \div 0.75\pi$
$= 212.20659\ldots$
Number of complete turns $= 212$

3 Work out the area of this compound shape.

Leave your answer in terms of π.

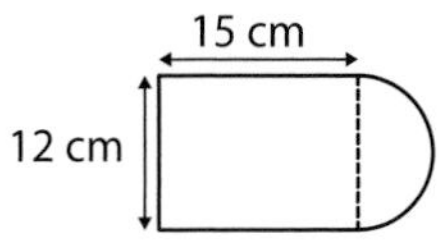

15 cm
12 cm

Area of rectangle $= 12 \times 15 = 180$
Area of semicircle $= \frac{1}{2} \times \pi \times 6^2 = 18\pi$
Area of compound shape $= 180 + 18\pi$

2 Checklist

- ✓ If the formula is in terms of radius r and the diameter is given, then divide the diameter by 2 to obtain the radius.
- ✓ Always draw lines to break up a compound shape into simple shapes.

Problem solving

1. Work out the circumference of the bicycle wheel.
2. Work out how many 'circumferences' go into 500 m.
3. Round down to find the number of complete turns.

10 Exam-style practice — Grades 5–6

The diagram shows a roller in the shape of a cylinder.

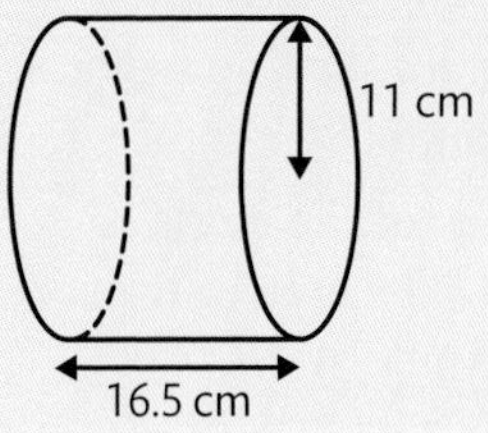

11 cm
16.5 cm

The roller has a radius of 11 cm and a length of 16.5 cm.
A company wants to make a new size of roller. The new roller will have a radius of 12.4 cm.
It will have the same volume as the large roller.

Work out the length of the new roller. Give your answer correct to one decimal place. **[3 marks]**

Circles, sectors and arcs

An arc is a fraction of the circumference of a circle and a sector is a 'slice' of the circle. Arc length and sector area are calculated as fractions of circumference or area, based on the angle at the centre.

5 Minor and major

A circle can be divided into two regions called the **minor sector** and the **major sector**.

The circumference of a circle can be divided into two parts called the **minor arc** and the **major arc**.

We often use θ for an unknown angle.

> The angle at the centre of a major arc or sector is greater than 180°.

> The angle at the centre of a minor arc or sector is less than 180°.

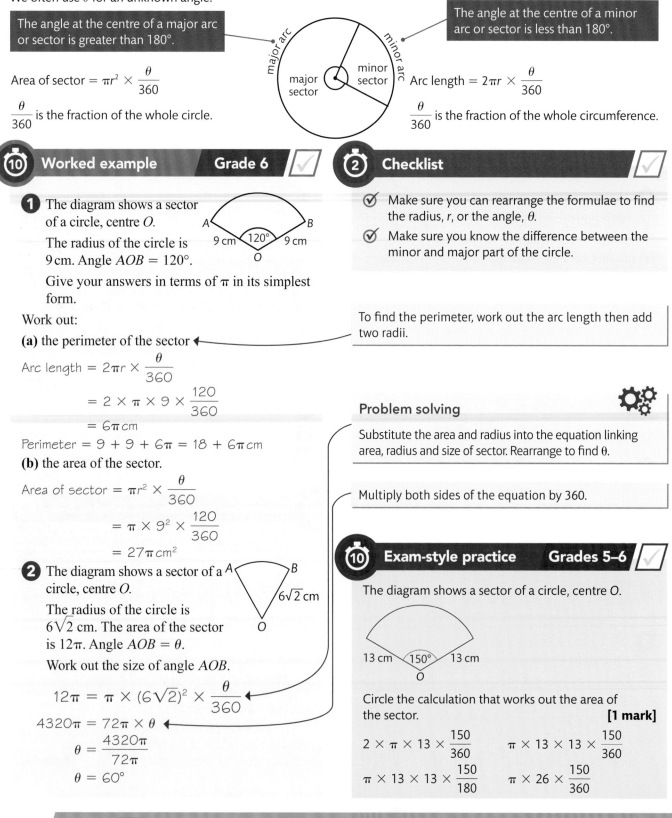

Area of sector $= \pi r^2 \times \dfrac{\theta}{360}$

$\dfrac{\theta}{360}$ is the fraction of the whole circle.

Arc length $= 2\pi r \times \dfrac{\theta}{360}$

$\dfrac{\theta}{360}$ is the fraction of the whole circumference.

10 Worked example — Grade 6

1 The diagram shows a sector of a circle, centre O.

The radius of the circle is 9 cm. Angle $AOB = 120°$.

A 9 cm 120° 9 cm *B*, *O*

Give your answers in terms of π in its simplest form.

Work out:

(a) the perimeter of the sector

Arc length $= 2\pi r \times \dfrac{\theta}{360}$

$= 2 \times \pi \times 9 \times \dfrac{120}{360}$

$= 6\pi$ cm

Perimeter $= 9 + 9 + 6\pi = 18 + 6\pi$ cm

(b) the area of the sector.

Area of sector $= \pi r^2 \times \dfrac{\theta}{360}$

$= \pi \times 9^2 \times \dfrac{120}{360}$

$= 27\pi$ cm^2

2 The diagram shows a sector of a circle, centre O.

A *B* $6\sqrt{2}$ cm *O*

The radius of the circle is $6\sqrt{2}$ cm. The area of the sector is 12π. Angle $AOB = \theta$.

Work out the size of angle AOB.

$12\pi = \pi \times (6\sqrt{2})^2 \times \dfrac{\theta}{360}$

$4320\pi = 72\pi \times \theta$

$\theta = \dfrac{4320\pi}{72\pi}$

$\theta = 60°$

2 Checklist

☑ Make sure you can rearrange the formulae to find the radius, r, or the angle, θ.

☑ Make sure you know the difference between the minor and major part of the circle.

To find the perimeter, work out the arc length then add two radii.

Problem solving

Substitute the area and radius into the equation linking area, radius and size of sector. Rearrange to find θ.

Multiply both sides of the equation by 360.

10 Exam-style practice — Grades 5–6

The diagram shows a sector of a circle, centre O.

13 cm 150° 13 cm *O*

Circle the calculation that works out the area of the sector. **[1 mark]**

$2 \times \pi \times 13 \times \dfrac{150}{360}$ $\qquad$ $\pi \times 13 \times 13 \times \dfrac{150}{360}$

$\pi \times 13 \times 13 \times \dfrac{150}{180}$ $\qquad$ $\pi \times 26 \times \dfrac{150}{360}$

Made a start ☑ Feeling confident ☑ Exam ready ☑

Circle facts

You will need to know the different properties of circles relating to chords, tangents, diameters and radii.

5 Properties of circles

Tangent and radius

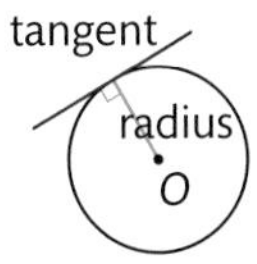

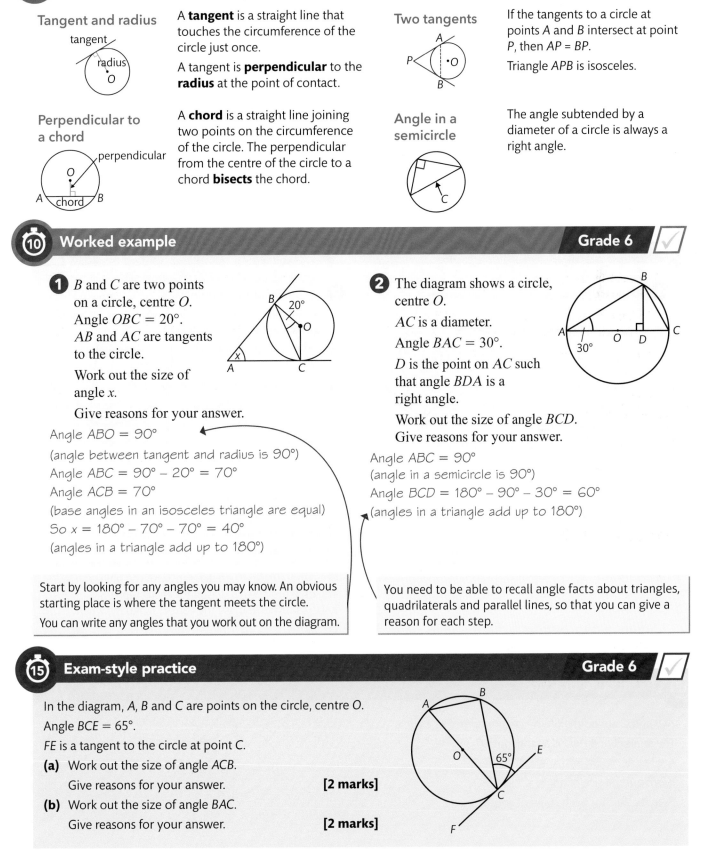

A **tangent** is a straight line that touches the circumference of the circle just once.

A tangent is **perpendicular** to the **radius** at the point of contact.

Two tangents

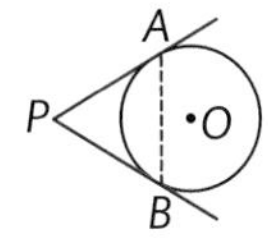

If the tangents to a circle at points *A* and *B* intersect at point *P*, then *AP* = *BP*.

Triangle *APB* is isosceles.

Perpendicular to a chord

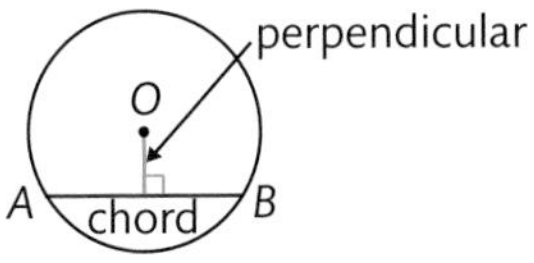

A **chord** is a straight line joining two points on the circumference of the circle. The perpendicular from the centre of the circle to a chord **bisects** the chord.

Angle in a semicircle

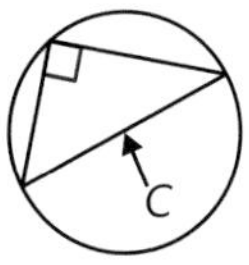

The angle subtended by a diameter of a circle is always a right angle.

10 Worked example — Grade 6

1 *B* and *C* are two points on a circle, centre *O*.
Angle $OBC = 20°$.
AB and *AC* are tangents to the circle.

Work out the size of angle *x*.

Give reasons for your answer.

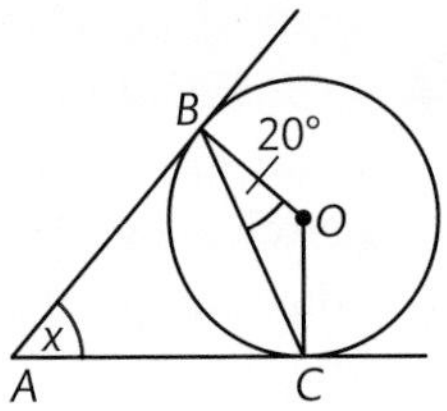

Angle $ABO = 90°$
(angle between tangent and radius is 90°)
Angle $ABC = 90° - 20° = 70°$
Angle $ACB = 70°$
(base angles in an isosceles triangle are equal)
So $x = 180° - 70° - 70° = 40°$
(angles in a triangle add up to 180°)

Start by looking for any angles you may know. An obvious starting place is where the tangent meets the circle.
You can write any angles that you work out on the diagram.

2 The diagram shows a circle, centre *O*.

AC is a diameter.

Angle $BAC = 30°$.

D is the point on *AC* such that angle *BDA* is a right angle.

Work out the size of angle *BCD*.
Give reasons for your answer.

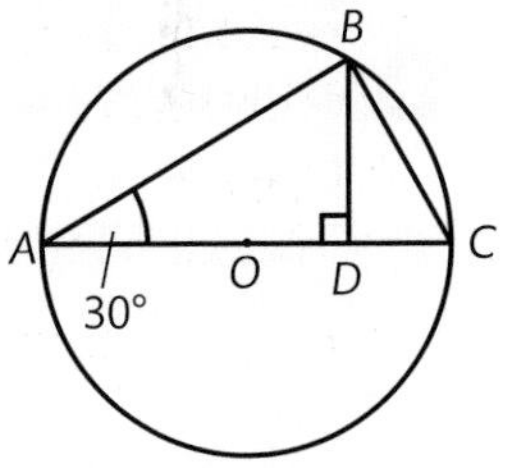

Angle $ABC = 90°$
(angle in a semicircle is 90°)
Angle $BCD = 180° - 90° - 30° = 60°$
(angles in a triangle add up to 180°)

You need to be able to recall angle facts about triangles, quadrilaterals and parallel lines, so that you can give a reason for each step.

15 Exam-style practice — Grade 6

In the diagram, *A*, *B* and *C* are points on the circle, centre *O*.
Angle $BCE = 65°$.
FE is a tangent to the circle at point *C*.

(a) Work out the size of angle *ACB*.
Give reasons for your answer. **[2 marks]**

(b) Work out the size of angle *BAC*.
Give reasons for your answer. **[2 marks]**

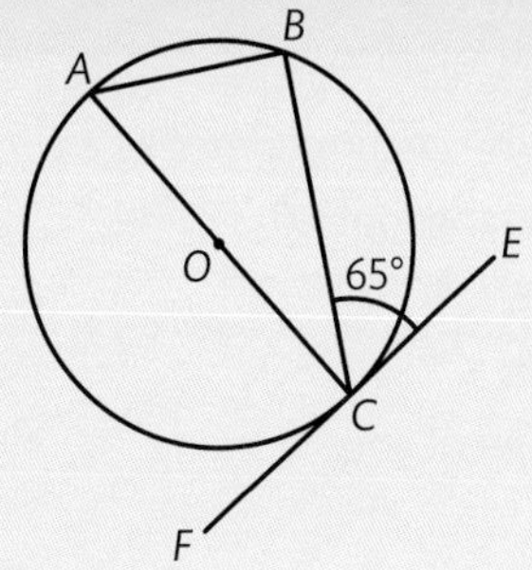

Made a start | Feeling confident | Exam ready

Circle theorems

You need to remember these circle theorems for the exam.

5 Circle theorems

You need to be able to quote and apply these theorems to circles.

Angles subtended by an arc at the centre and circumference

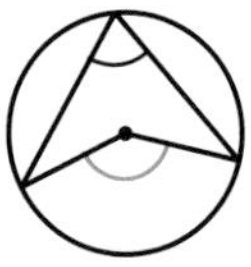

The angle subtended at the centre is twice the angle subtended at the circumference.

Opposite angles in a cyclic quadrilateral

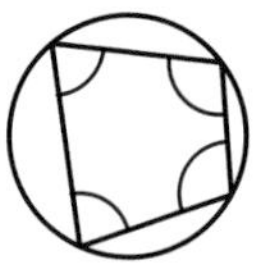

Opposite angles in a cyclic quadrilateral add up to 180°.

Angles subtended in the same segment

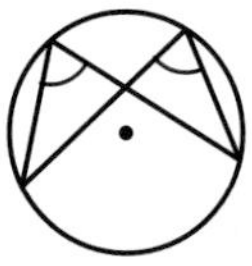

Angles in the same segment are equal.

Alternate segment theorem

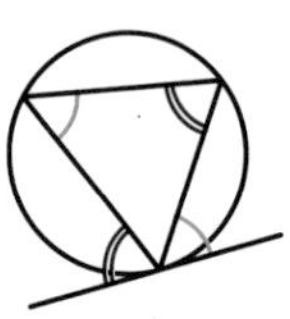

The angle between a tangent and a chord is equal to the angle subtended in the alternate segment.

10 Worked example — Grade 7

1 In the diagram, A, B, C and D are points on the circumference of a circle, centre O.

Angle $BAD = 65°$

Angle $BOD = x$

Angle $BCD = y$

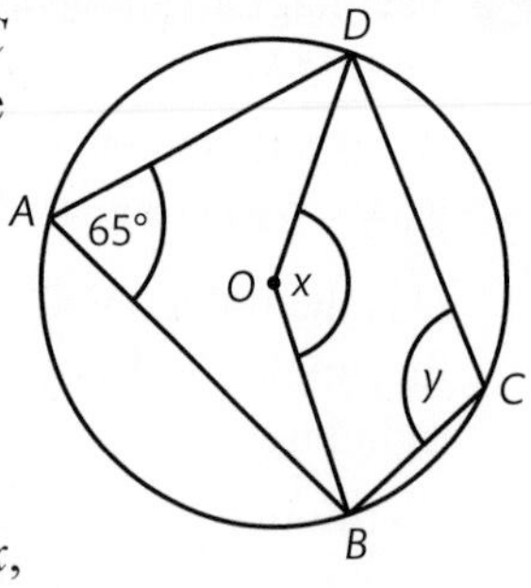

(a) Work out the value of x, giving a reason for your answer.

$x = 2 \times 65° = 130°$

(angle at centre is twice angle at circumference)

(b) Work out the value of y, giving a reason for your answer.

$y = 180° - 65° = 115°$

(opposite angles in a cyclic quadrilateral add up to 180°)

2 A, B, C and D are points on the circumference of a circle, centre O. AC is a diameter of the circle. Angle $DAC = 25°$.

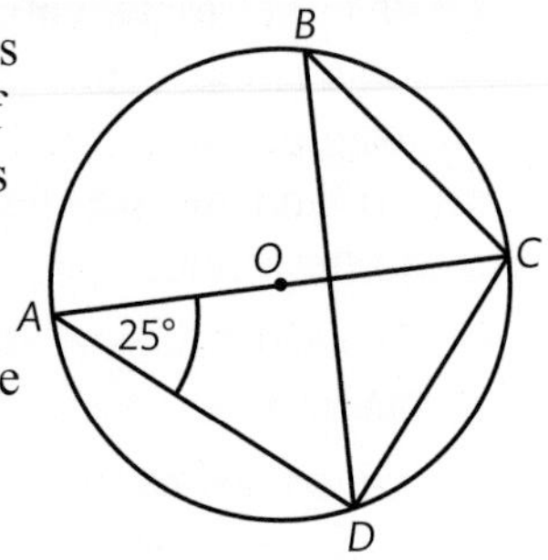

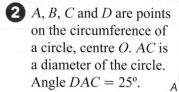

(a) Work out the size of angle ADC, giving a reason for your answer.

angle $ADC = 90°$

(angle in a semicircle is a right angle)

(b) Work out the size of angle DBC, giving a reason for your answer.

angle $DBC = 25°$

(angles in the same segment are equal)

When you are answering circle theorem questions, you can annotate the given diagram with the unknown angles. Write down a reason for each step. It is important that you use the correct language when giving reasons.

15 Exam-style practice — Grade 7

A, B and C are points on the circumference of a circle.

The straight line PAQ is a tangent to the circle.

Angle $PAC = 58°$ and angle $ACB = 76°$.

Work out the size of angle x.

Give reasons for each stage of your working.

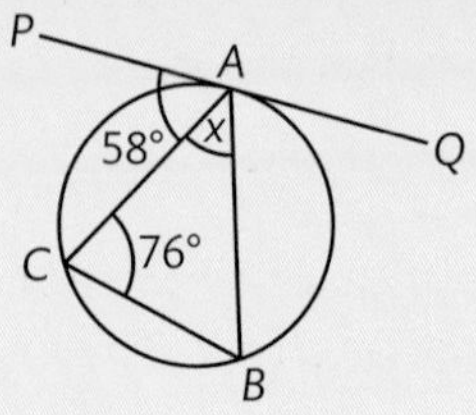

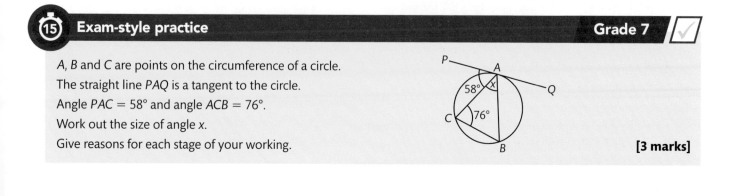

[3 marks]

Transformations

A transformation changes the position, size or orientation of a shape.

 5 Types of transformation

Here are three out of the four transformations you need to know about. The original shape is the **object** and the final shape is its **image**.

Reflection	Rotation	Translation

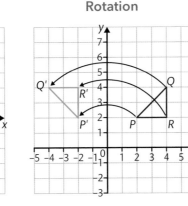

A **reflection** in a line produces another shape that is a mirror image of the original shape. The image is always the same distance from the mirror line as the object was. You can describe a reflection by giving the equation of its mirror line.

A **rotation** turns a shape through a clockwise or anticlockwise direction about a fixed point. The fixed point is the **centre of rotation**. Rotation changes the position and orientation of the shape.

A **translation** changes the position of the shape. It moves the object horizontally and/or vertically, but does not change its orientation. It moves all its points the same distance in the same direction. A translation can be described by using vector notation, $\begin{pmatrix} x \\ y \end{pmatrix}$.

---

10 Worked example — Grade 4

1

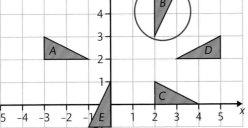

(a) Triangle A is rotated 90° clockwise about the centre (1, 1). Circle the triangle that this maps to: B, C, D or E.

(b) Describe fully the single transformation that maps triangle A onto triangle C.

Translation $\begin{pmatrix} 5 \\ -2 \end{pmatrix}$

(c) Describe fully the single transformation that maps triangle A onto triangle D.

Reflection in the line $x = 1$

---

15 Exam-style practice — Grade 4

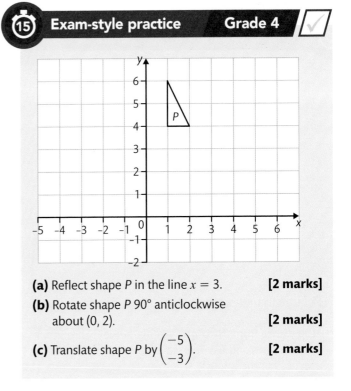

(a) Reflect shape P in the line $x = 3$. [2 marks]

(b) Rotate shape P 90° anticlockwise about (0, 2). [2 marks]

(c) Translate shape P by $\begin{pmatrix} -5 \\ -3 \end{pmatrix}$. [2 marks]

✓ **Made a start** ✓ **Feeling confident** ✓ **Exam ready**

Enlargement

The **enlargement** transformation makes the shape larger or smaller, depending on the **scale factor**. Every enlargement takes place from a **centre of enlargement**. There are two types of enlargement: those with **positive scale factors** and those with **negative scale factors**.

⑤ Positive scale factor ✓

For a scale factor greater than 1, the image is larger than the original shape.
Scale factor for image A'B'C' = 2

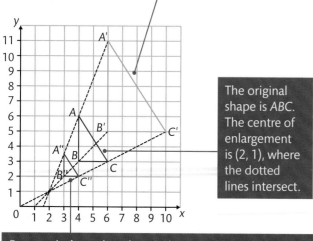

The original shape is ABC. The centre of enlargement is (2, 1), where the dotted lines intersect.

For a scale factor less than 1, the image is smaller than the original shape.
Scale factor for image A"B"C" = $\frac{1}{2}$

⑤ Negative scale factor ✓

A **negative** scale factor produces an image that is on the opposite side of the centre of enlargement from the original shape. It rotates it through 180°.

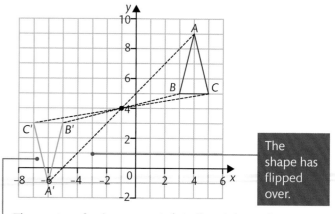

The shape has flipped over.

The centre of enlargement is (–1, 4) and the scale factor is –1.

The original shape is ABC. For image A'B'C', the centre of enlargement is (–1, 4) and the scale factor is -1.

For a negative number less than –1, the image will be larger than the original shape.

For a negative decimal or fraction, the image will be smaller than the shape.

② Checklist ✓

- ☑ Make sure you know the difference between positive scale factors and negative scale factors.
- ☑ A negative scale factor will flip the shape over.
- ☑ To describe an enlargement, write down the scale factor and centre of enlargement.

⑤ Worked example — Grade 4 ✓

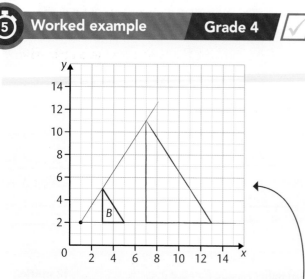

Enlarge triangle B by scale factor 3, centre (1, 2).

Draw lines from the centre of enlargement through the vertices, and use the scale factor to locate the vertices of the image.

⑩ Exam-style practice — Grade 5 ✓

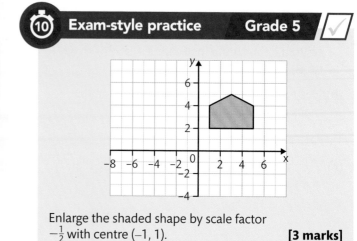

Enlarge the shaded shape by scale factor $-\frac{1}{2}$ with centre (–1, 1). **[3 marks]**

✓ **Made a start** ✓ **Feeling confident** ✓ **Exam ready**

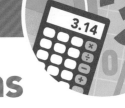

Combining transformations

Sometimes you can describe the result of two or more transformations as a single transformation. Objects that stay the same (do not change) are **invariant**.

5 Summary of transformations

Transformation	How to describe it	What does not change?	What can change?
enlargement	centre of enlargement scale factor	angles, ratios of side lengths	size, position and orientation
reflection	equation of line of reflection	angles, shape and size	position and orientation
rotation	centre of rotation clockwise or anticlockwise size of angle	angles, shape and size	position and orientation
translation	translation vector	angles, shape, size and orientation	position

5 Worked example — Grade 6

1

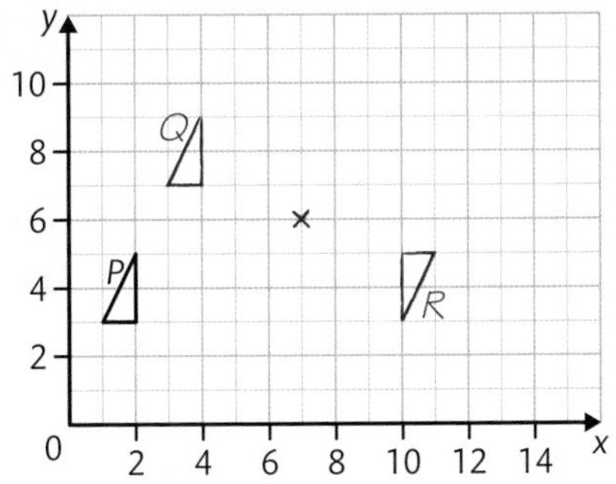

(a) Translate shape P by the vector $\begin{pmatrix} 2 \\ 4 \end{pmatrix}$.

Label the new shape Q.

(b) Rotate shape Q 180° about (7, 6).

Label the new shape R.

(c) Describe fully the single transformation that maps shape P onto shape R.

Rotation 180° about (6, 4).

2

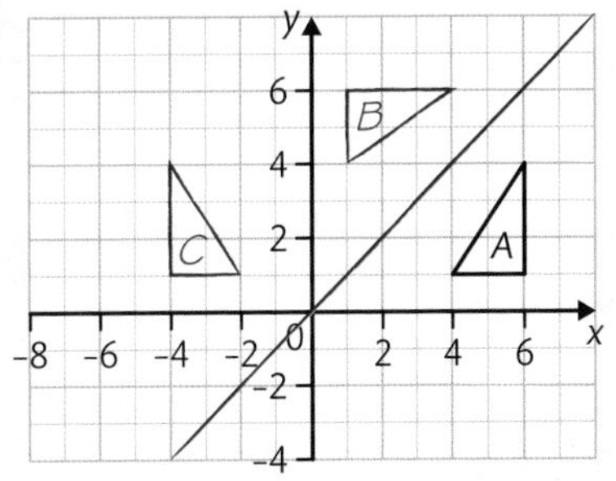

(a) Reflect triangle A in the line $y = x$.

Label the new shape B.

(b) Rotate shape B 90° anticlockwise about (1, 1).

Label the new shape C.

(c) Describe fully the single transformation that will map shape A onto shape C.

Reflection in the line $x = 1$

Problem solving

You can use tracing paper to help you find a centre of rotation. Trace the original shape and place your pencil or compass points at different coordinates. You need to find the point around which you can rotate the tracing paper so the object matches the image.

15 Exam-style practice — Grade 6

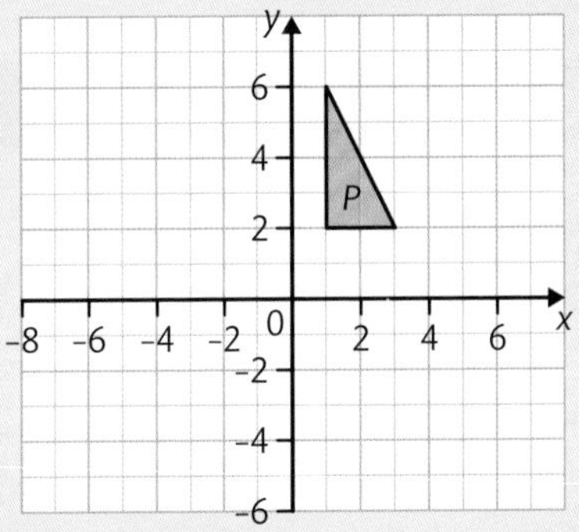

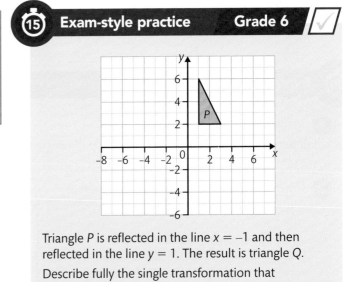

Triangle P is reflected in the line $x = -1$ and then reflected in the line $y = 1$. The result is triangle Q.

Describe fully the single transformation that will map triangle P onto triangle Q. **[3 marks]**

Bearings

A **bearing** is a measurement of the position of one point relative to another point. You will also see bearings described in terms of *N* (north), *S* (south), *E* (east) and *W* (west).

2 Language of bearings

Bearings are always measured from the **north** in a **clockwise** direction.

Bearings are measured in degrees.

Bearings are given as three digits.

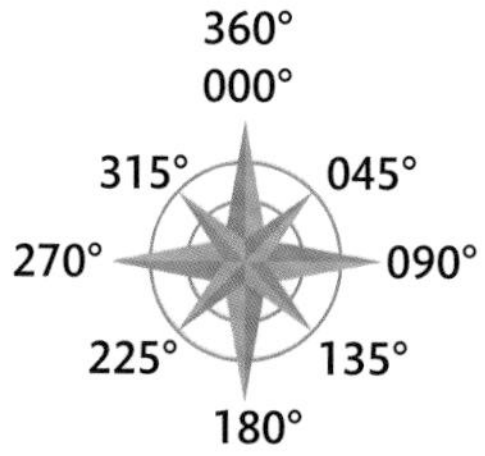
360°
000°
315°
045°
270°
090°
225°
135°
180°

Make sure you know the meaning of 'bearing of *B* from *F*'. This means you measure from *F*.

5 Worked example — Grade 5

(a) Write down the bearing of B from F.

Bearing of B from F = 075°

(b) Write down the bearing of F from B.

Bearing of F from B = 180° + 75° = 255°

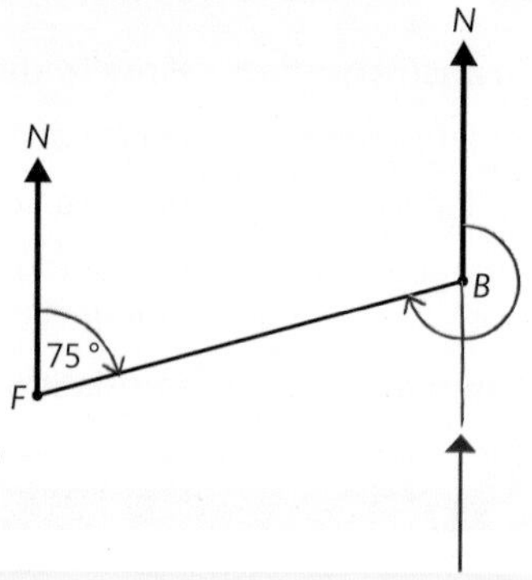
N
N
B
75°
F

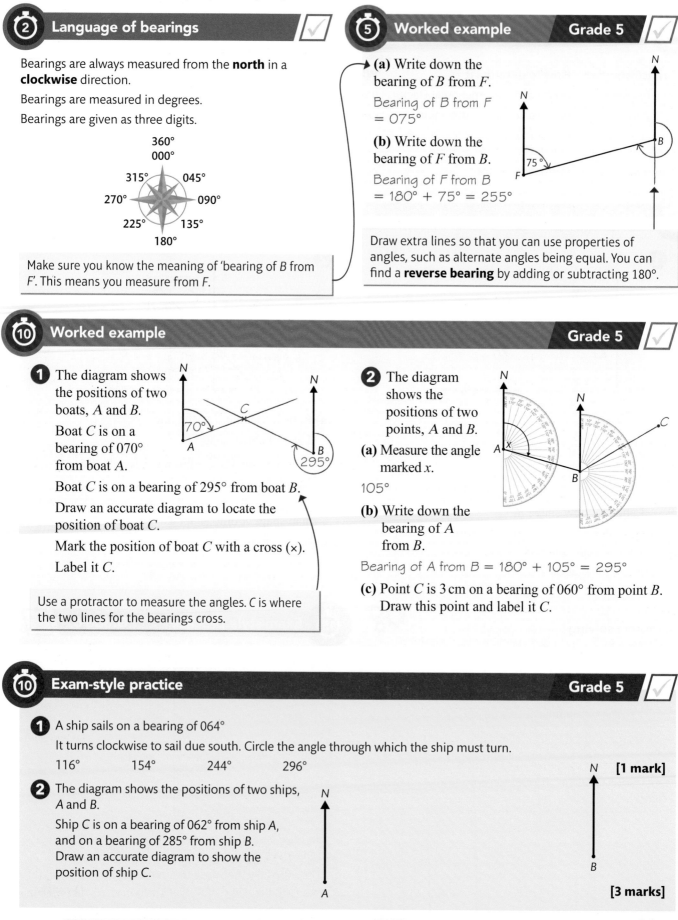

Draw extra lines so that you can use properties of angles, such as alternate angles being equal. You can find a **reverse bearing** by adding or subtracting 180°.

10 Worked example — Grade 5

1 The diagram shows the positions of two boats, A and B.

Boat C is on a bearing of 070° from boat A.

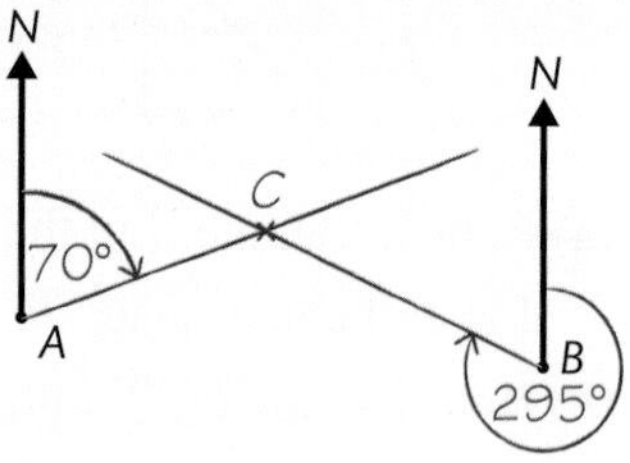
N
N
C
70°
A
B
295°

Boat C is on a bearing of 295° from boat B.

Draw an accurate diagram to locate the position of boat C.

Mark the position of boat C with a cross (×).

Label it C.

Use a protractor to measure the angles. *C* is where the two lines for the bearings cross.

2 The diagram shows the positions of two points, A and B.

(a) Measure the angle marked x.

105°

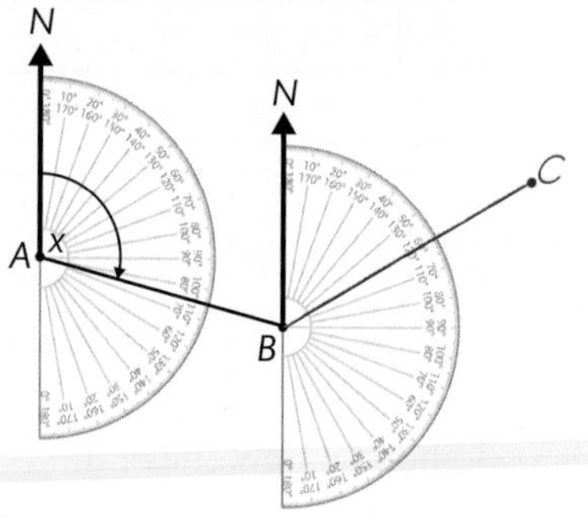
N
N
A
X
B
C

(b) Write down the bearing of A from B.

Bearing of A from B = 180° + 105° = 295°

(c) Point C is 3 cm on a bearing of 060° from point B. Draw this point and label it C.

10 Exam-style practice — Grade 5

1 A ship sails on a bearing of 064°

It turns clockwise to sail due south. Circle the angle through which the ship must turn.

116° 154° 244° 296° **[1 mark]**

2 The diagram shows the positions of two ships, *A* and *B*.

Ship C is on a bearing of 062° from ship *A*, and on a bearing of 285° from ship *B*.
Draw an accurate diagram to show the position of ship C.

N
A

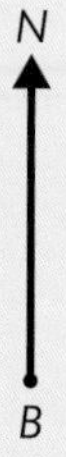
N
B

[3 marks]

Scale drawings and maps

You can use ratios to work out problems involving scale drawings and map scales.

⑤ Scale drawings ✓

In a scale drawing, all the dimensions are reduced by the same proportion.

You will need to know how to read a scale. It may be written in the form 1 cm = 30 m or as a ratio, in the form 1 : 3000.

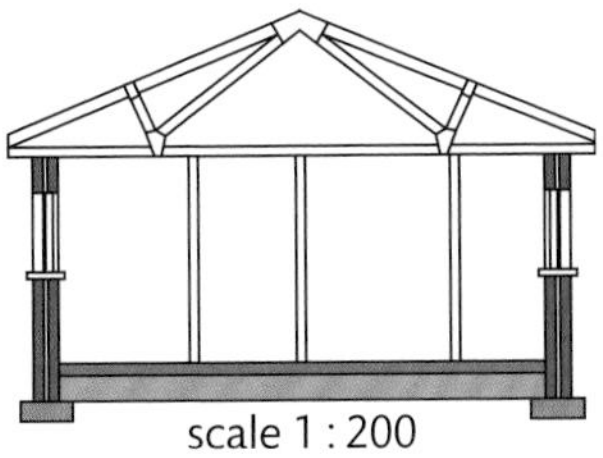

scale 1 : 200

What does a scale of 1 : 25 mean?

This scale can be applied to any units. Using this scale, 1 mm would represent 25 mm or 2.5 cm, 1 cm would represent 25 cm or 0.25 m, and so on.

⑤ Maps ✓

Map scales may be written in the form 1 cm = 1 km or as ratios, such as 1 : 100 000. They tell you how real-life measurements are represented on a map.

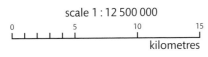

scale 1 : 12 500 000

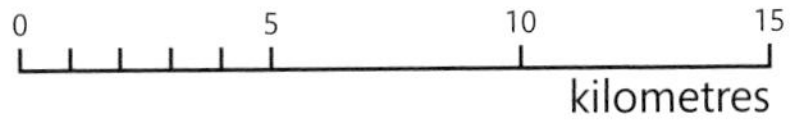

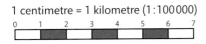

1 centimetre = 1 kilometre (1 : 100 000)

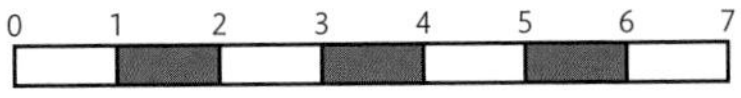

What does a map scale of 1 : 25 000 mean?

If a distance is measured as 5 cm on the map, the real distance can be worked out:

5 cm represents 5 × 25 000 = 125 000 cm

It is not sensible to write a distance as 125 000 cm.

÷100 ÷1000

125 000 cm = 1250 m = 1.25 km

② Checklist ✓

- ✓ Make sure you can interpret scales such as 1 : *n*.
- ✓ Make sure you measure the map distance carefully.
- ✓ Make sure that the scale on the ruler begins at 0.
- ✓ Make sure you can recall conversion factors.
- ✓ Always give an answer in sensible units.

⑤ Worked example — Grade 5 ✓

The length of a van is 4.2 metres. Alice makes a scale model of the van. She uses a scale of 1 : 40.

Work out the length of the scale model of the van.

Give your answer in centimetres.

Length of van is 4.2 × 100 = 420 cm

Length of scale model is 420 ÷ 40 = 10.5 cm

Divide by 40, because every 40 cm on the van will be represented by 1 cm in the model.

⑤ Worked example — Grade 5 ✓

1 The diagram shows the positions of two hills, Penn and Hagley, on a map.

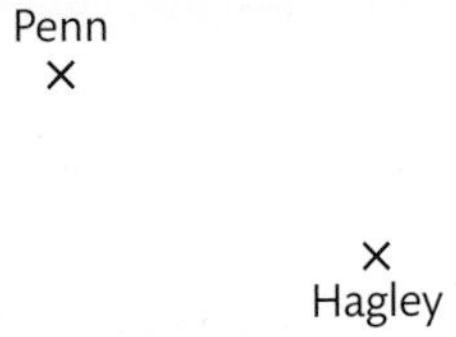

The scale of the map is 1 cm = 3.5 km. Work out the real distance between Penn and Hagley.

Map distance = 2.5 cm

Actual distance = 2.5 × 3.5 = 8.75 km

2 The scale of a map is 1 : 25 000. On the map, the distance between two villages is 21 cm. Work out the real distance between the two villages.

Give your answer in kilometres.

Actual distance = 21 × 25 000 = 525 000 cm
= 5250 m
= 5.25 km

⑤ Exam-style practice — Grade 5 ✓

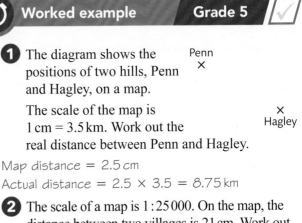

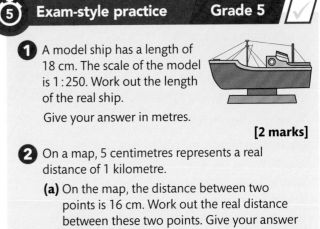

1 A model ship has a length of 18 cm. The scale of the model is 1 : 250. Work out the length of the real ship.

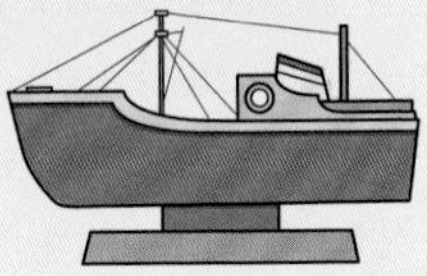

Give your answer in metres.

[2 marks]

2 On a map, 5 centimetres represents a real distance of 1 kilometre.

(a) On the map, the distance between two points is 16 cm. Work out the real distance between these two points. Give your answer in kilometres. **[2 marks]**

(b) Circle the correct map scale

1 : 200 1 : 2000 1 : 20 000 1 : 200 000

[1 mark]

Similar shapes

If two shapes are similar, you can use scale factors to compare their lengths, areas and volumes.

(5) Understanding scale factors

The diagram shows two solid cuboids that are mathematically similar. Shape B is an enlargement of shape A with a linear scale factor of k.

Scale factor from	Length	Area	Volume
A to B	$\times k$	$\times k^2$	$\times k^3$
B to A	$\div k$	$\div k^2$	$\div k^3$

(5) Worked example   Grade 7

Here are two shop price tickets.

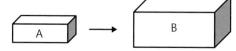

The two shop price tickets are mathematically similar.

(a) Work out the length x.

Scale factor = $7.5 \div 3 = 2.5$

$x = 5 \times 2.5 = 12.5\,$cm

The area of the larger ticket is $75\,$cm^2.

(b) Work out the area of the smaller ticket.

Area of smaller ticket = $75 \div 2.5^2 = 12\,$cm^2

(5) Worked example   Grade 8

X and Y are two similar 3-D shapes.

The surface area of A is $6000\,$cm^2.

The surface area of B is $13\,500\,$cm^2.

The volume of B is $37\,125\,$cm^3.

Work out the volume of A.

Area scale factor = $13\,500 \div 6000 = 2.25$
Length scale factor = $\sqrt{2.25} = 1.5$
Volume scale factor = 1.5^3
Volume of shape A = $37\,125 \div 1.5^3$
Volume of shape A = $11\,000\,$cm^3

To find the length scale factor, you need to square root the area scale factor or cube root the volume scale factor.

(5) Similar triangles

You need to be able to recognise the conditions under which two triangles are **similar**:

1 all three sides in the same proportion

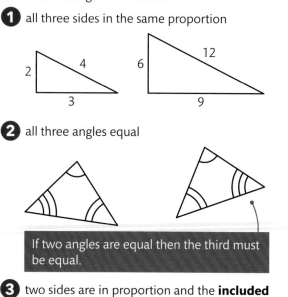

2 all three angles equal

If two angles are equal then the third must be equal.

3 two sides are in proportion and the **included** angle is equal.

Watch out, the equal angle must be **between** the two sides.

(10) Exam-style practice   Grades 7–8

1 Cylinders A and B are mathematically similar.
A has height $4\,$cm and base diameter $7\,$cm.
B has height $14\,$cm and base diameter $d\,$cm.
(a) Work out the value of d.   **[2 marks]**
The surface area of A is $165\,$cm^2.
(b) Work out the surface area of B.   **[2 marks]**
The volume of B is $6602.75\,$cm^3
(c) Work out the volume of A.   **[2 marks]**

2
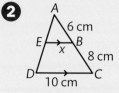

(a) Explain why triangles AEB and ADC are similar.   **[2 marks]**
(b) Work out the length EB, marked x on the diagram.   **[3 marks]**

Congruent triangles

Congruent shapes are identical in size and shape although their position relative to each other may be different.

5 Conditions for congruency

Two triangles are congruent if they both satisfy any one set of the following conditions.

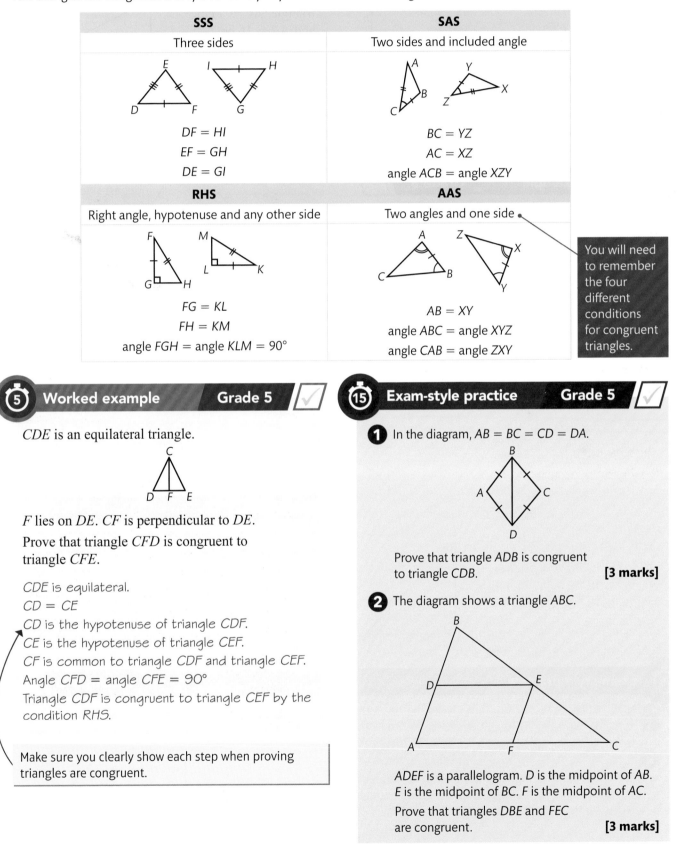

SSS	SAS
Three sides	Two sides and included angle
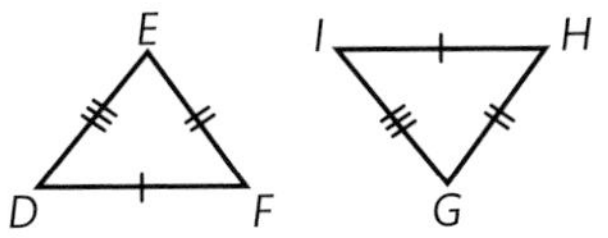	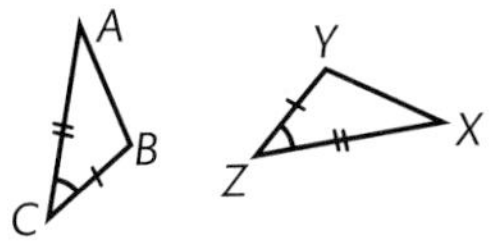
$DF = HI$ $EF = GH$ $DE = GI$	$BC = YZ$ $AC = XZ$ angle ACB = angle XZY
RHS	**AAS**
Right angle, hypotenuse and any other side	Two angles and one side
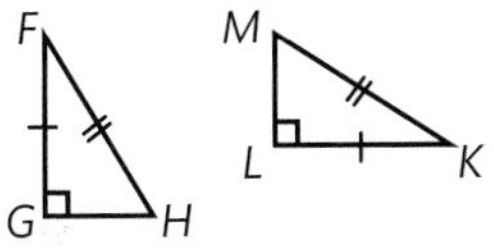	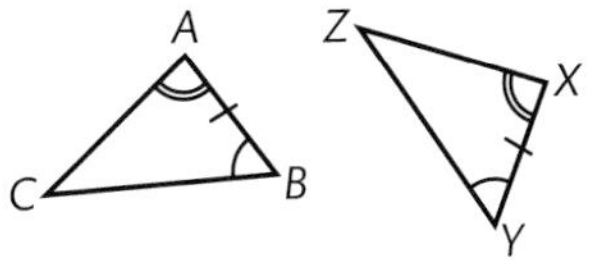
$FG = KL$ $FH = KM$ angle FGH = angle $KLM = 90°$	$AB = XY$ angle ABC = angle XYZ angle CAB = angle ZXY

You will need to remember the four different conditions for congruent triangles.

5 Worked example Grade 5

CDE is an equilateral triangle.

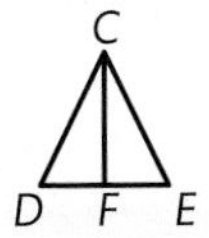

F lies on DE. CF is perpendicular to DE.

Prove that triangle CFD is congruent to triangle CFE.

CDE is equilateral.
$CD = CE$
CD is the hypotenuse of triangle CDF.
CE is the hypotenuse of triangle CEF.
CF is common to triangle CDF and triangle CEF.
Angle CFD = angle $CFE = 90°$
Triangle CDF is congruent to triangle CEF by the condition RHS.

Make sure you clearly show each step when proving triangles are congruent.

15 Exam-style practice Grade 5

1 In the diagram, $AB = BC = CD = DA$.

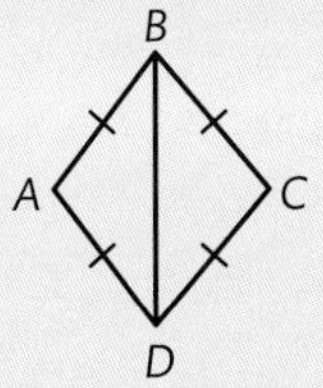

Prove that triangle ADB is congruent to triangle CDB. **[3 marks]**

2 The diagram shows a triangle ABC.

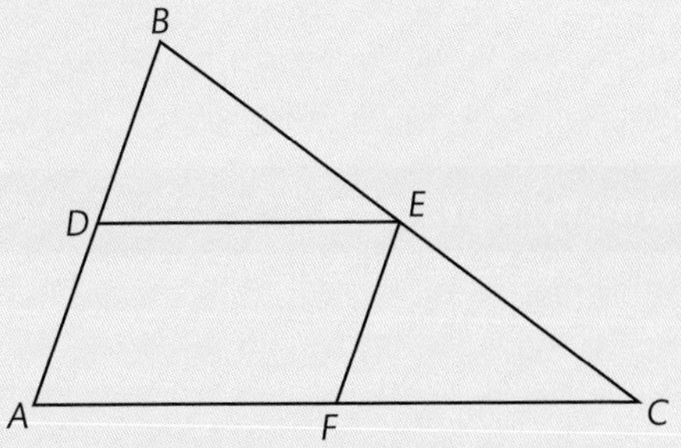

$ADEF$ is a parallelogram. D is the midpoint of AB. E is the midpoint of BC. F is the midpoint of AC.

Prove that triangles DBE and FEC are congruent. **[3 marks]**

Pythagoras' theorem

If you know two sides of a right-angled triangle, you can use Pythagoras' theorem to work out the length of the third side.

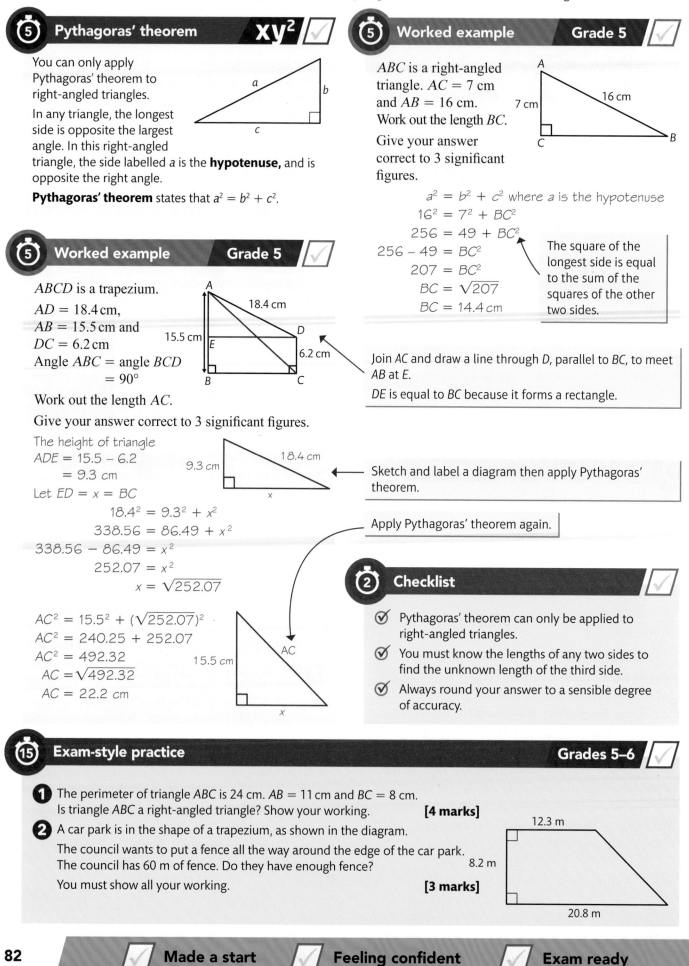

(5) Pythagoras' theorem xy² ✓

You can only apply Pythagoras' theorem to right-angled triangles.

In any triangle, the longest side is opposite the largest angle. In this right-angled triangle, the side labelled a is the **hypotenuse,** and is opposite the right angle.

Pythagoras' theorem states that $a^2 = b^2 + c^2$.

(5) Worked example Grade 5 ✓

$ABCD$ is a trapezium.

$AD = 18.4\,cm$,
$AB = 15.5\,cm$ and
$DC = 6.2\,cm$
Angle ABC = angle BCD
    $= 90°$

Work out the length AC.

Give your answer correct to 3 significant figures.

The height of triangle
$ADE = 15.5 - 6.2$
    $= 9.3\ cm$
Let $ED = x = BC$

$18.4^2 = 9.3^2 + x^2$
$338.56 = 86.49 + x^2$
$338.56 - 86.49 = x^2$
$252.07 = x^2$
$x = \sqrt{252.07}$

$AC^2 = 15.5^2 + (\sqrt{252.07})^2$
$AC^2 = 240.25 + 252.07$
$AC^2 = 492.32$
$AC = \sqrt{492.32}$
$AC = 22.2\ cm$

(5) Worked example Grade 5 ✓

ABC is a right-angled triangle. $AC = 7$ cm and $AB = 16$ cm. Work out the length BC.

Give your answer correct to 3 significant figures.

$a^2 = b^2 + c^2$ where a is the hypotenuse
$16^2 = 7^2 + BC^2$
$256 = 49 + BC^2$
$256 - 49 = BC^2$
$207 = BC^2$
$BC = \sqrt{207}$
$BC = 14.4\ cm$

The square of the longest side is equal to the sum of the squares of the other two sides.

Join AC and draw a line through D, parallel to BC, to meet AB at E.

DE is equal to BC because it forms a rectangle.

Sketch and label a diagram then apply Pythagoras' theorem.

Apply Pythagoras' theorem again.

(2) Checklist ✓

☑ Pythagoras' theorem can only be applied to right-angled triangles.

☑ You must know the lengths of any two sides to find the unknown length of the third side.

☑ Always round your answer to a sensible degree of accuracy.

(15) Exam-style practice Grades 5–6 ✓

1 The perimeter of triangle ABC is 24 cm. $AB = 11$ cm and $BC = 8$ cm.
Is triangle ABC a right-angled triangle? Show your working. **[4 marks]**

2 A car park is in the shape of a trapezium, as shown in the diagram.

The council wants to put a fence all the way around the edge of the car park.
The council has 60 m of fence. Do they have enough fence?

You must show all your working. **[3 marks]**

12.3 m

8.2 m

20.8 m

✓ **Made a start**   ✓ **Feeling confident**   ✓ **Exam ready**

Pythagoras' theorem in 3D

You can use Pythagoras' theorem to find unknown lengths in three dimensions.

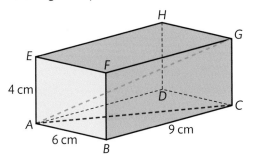

3D problems

Pythagoras' theorem can be applied to cubes, cuboids, cones and pyramids.

This diagram represents a cuboid *ABCDEFGH*.

How can you work out the length of *AG*, correct to 3 significant figures?

Step 1: Apply Pythagoras' theorem to work out *AC*, which is in triangle *ABC* and in triangle *CAG*.

$AC^2 = 6^2 + 9^2$

$AC^2 = 36 + 81$

$AC^2 = 117$

$AC = \sqrt{117}$

Step 2: Apply Pythagoras' theorem to work out *AG*, which is in triangle *CAG*.

$AG^2 = (\sqrt{117})^2 + 4^2$

$AG^2 = 117 + 16$

$AG^2 = 133$

$AG = \sqrt{133}$

$AG = 11.5$ cm

> Do not round any values until your final answer.

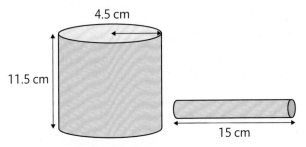

Worked example — Grade 7

The diagram shows a cylindrical container and a narrow steel rod. The height of the cylinder is 11.5 cm and its radius is 4.5 cm. The length of the steel rod is 15 cm.

Can the rod fit inside the container? You must show all your working.

$x^2 = 11.5^2 + 9^2$

$x^2 = 132.25 + 81$

$x^2 = 213.25$

$x = \sqrt{213.25}$

$x = 14.60308...$ cm

As 14.6 cm < 15 cm the rod cannot fit inside the container.

> Draw a right-angled triangle to work out the hypotenuse of a triangle of height 11.5 cm and base 2 × 4.5 cm.
>
>
>
> This is the maximum internal dimension of the cylinder.

> To answer the question you must make a statement at the end.

Checklist

- ☑ Make sure you know Pythagoras' theorem and how to use it.
- ☑ Always draw extra lines so that you know how to work out each step.
- ☑ Try to plan your method and your answer.
- ☑ Remember that the greatest dimension of a cuboid is the longest diagonal, between opposite vertices.

Exam-style practice — Grade 7

VABCD is a right pyramid on a square base. *V* is vertically above the centre of the square.

$VA = VB = VC = VD = 30$ cm

$AB = 22$ cm

Work out the vertical height of the pyramid.

[4 marks]

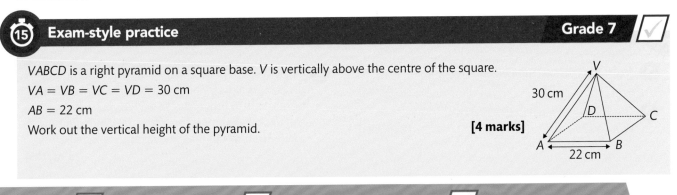

Units of length, area and volume

You should be able to convert between different units for lengths, areas and volumes.

⏱ 5 Conversions ☑

Converting units of length

$1\,cm = 10\,mm$

Converting units of area

For area the conversion factor is the **square** of the **length factor**.

$1\,cm^2 = 100\,mm^2$

Converting units of volume

For volume the conversion factor is the **cube** of the length factor.

> This is area so the scale factor for m^2 to cm^2 is 100 **squared**.
> m^2 to cm^2 is $\times 100^2$

⏱ 5 Worked example | Grade 5 ☑

1 Circle the area that is the same as $7.5\,m^2$.

$750\,cm^2$ $7500\,cm^2$ (75 000 cm²) $75\,000\,000\,cm^2$

$7.5 \times 100^2 = 75\,000\,cm^2$

2 Change $635\,259.7\,cm^3$ to m^3.

$635\,259.7 \div 100^3 = 0.635\,2597\,m^3$

3 Change 4.8 cubic metres to cubic centimetres.

$4.8\ cubic\ metres = 4.8\,m^3$
$= 4.8 \times 100^3$
$= 4\,800\,000\ cubic\ centimetres$

> This is volume so the scale factor for cm^3 to m^3 is 100^3.
> cm^3 to m^3 is $\div 100^3$ m^3 to cm^3 is $\times 100^3$

⏱ 10 Worked example | Grade 5 ☑

Ben and Nick went to London in their car. Ben said the speed of the car was 130 kilometres per hour. Nick told Ben that 130 kilometres per hour was about the same as 36 metres per second.

Was Nick correct? Show your working to justify your answer.

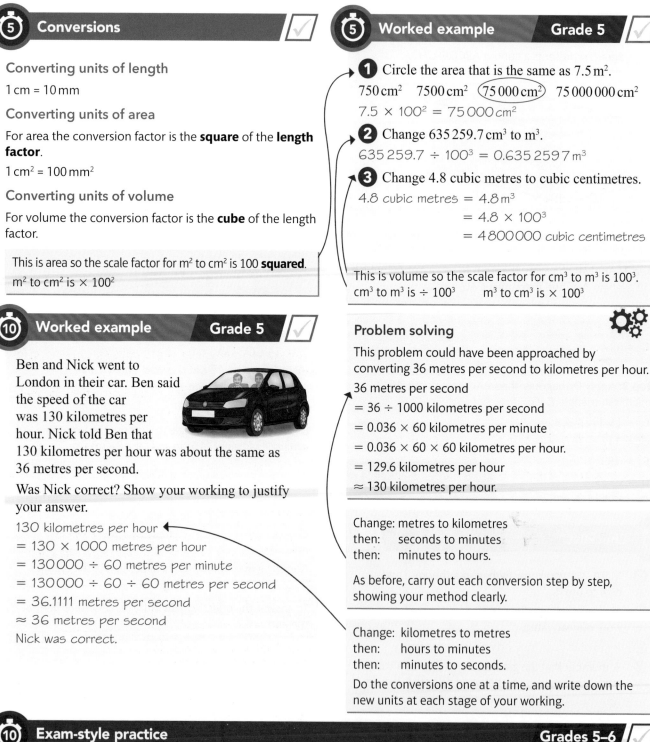

130 kilometres per hour
$= 130 \times 1000$ metres per hour
$= 130\,000 \div 60$ metres per minute
$= 130\,000 \div 60 \div 60$ metres per second
$= 36.1111$ metres per second
≈ 36 metres per second
Nick was correct.

⚙ Problem solving

This problem could have been approached by converting 36 metres per second to kilometres per hour.

36 metres per second
$= 36 \div 1000$ kilometres per second
$= 0.036 \times 60$ kilometres per minute
$= 0.036 \times 60 \times 60$ kilometres per hour.
$= 129.6$ kilometres per hour
≈ 130 kilometres per hour.

> Change: metres to kilometres
> then: seconds to minutes
> then: minutes to hours.
>
> As before, carry out each conversion step by step, showing your method clearly.

> Change: kilometres to metres
> then: hours to minutes
> then: minutes to seconds.
>
> Do the conversions one at a time, and write down the new units at each stage of your working.

⏱ 10 Exam-style practice | Grades 5–6 ☑

1 A rectangle has an area of $4\,m^2$. Write this area in cm^2. **[2 marks]**

2 The volume of a tile is $39\,250\,mm^3$. Write $39\,250\,mm^3$ in cm^3. **[2 marks]**

3 Singh Systems is a company that makes computer processors. For shipping, processors are packed into boxes with 30 processors in each box. A machine can pack 2880 processors per hour. Work out how many seconds the machine takes to fill one box. **[3 marks]**

☑ **Made a start**   ☑ **Feeling confident**   ☑ **Exam ready**

Trigonometry: lengths

You can use trigonometry to find the lengths of unknown sides in right-angled triangles.

(5) Sine, cosine and tangent xy^2 ✓

You need to learn the formulae for sine (sin), cosine (cos) and tangent (tan). These formulae work for any **right-angled triangle**. The sides of the triangle are labelled relative to one of the acute angles.

$$\sin \theta = \frac{\text{opp}}{\text{hyp}} \quad \text{SOH}$$

$$\cos \theta = \frac{\text{adj}}{\text{hyp}} \quad \text{CAH}$$

$$\tan \theta = \frac{\text{opp}}{\text{adj}} \quad \text{TOA}$$

Exam focus 📌

You need to learn the three trigonometric ratios for your exam. You can use SOH CAH TOA to help you remember them.

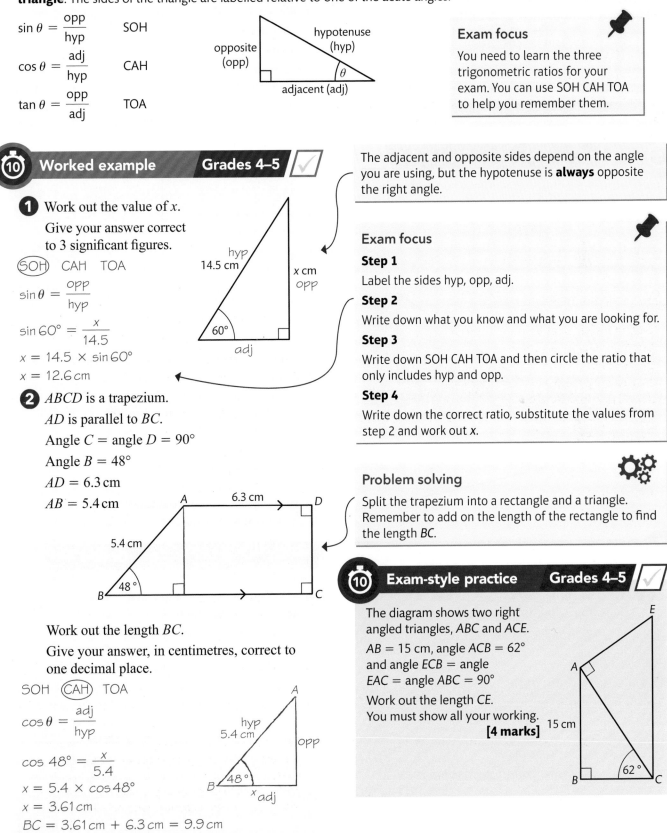

(10) Worked example Grades 4–5 ✓

1 Work out the value of x.

Give your answer correct to 3 significant figures.

(SOH) CAH TOA

$$\sin \theta = \frac{\text{opp}}{\text{hyp}}$$

$$\sin 60° = \frac{x}{14.5}$$

$$x = 14.5 \times \sin 60°$$

$$x = 12.6 \,\text{cm}$$

2 $ABCD$ is a trapezium.

AD is parallel to BC.

Angle C = angle D = 90°

Angle B = 48°

AD = 6.3 cm

AB = 5.4 cm

Work out the length BC.

Give your answer, in centimetres, correct to one decimal place.

SOH (CAH) TOA

$$\cos \theta = \frac{\text{adj}}{\text{hyp}}$$

$$\cos 48° = \frac{x}{5.4}$$

$$x = 5.4 \times \cos 48°$$

$$x = 3.61 \,\text{cm}$$

$$BC = 3.61 \,\text{cm} + 6.3 \,\text{cm} = 9.9 \,\text{cm}$$

The adjacent and opposite sides depend on the angle you are using, but the hypotenuse is **always** opposite the right angle.

Exam focus 📌

Step 1

Label the sides hyp, opp, adj.

Step 2

Write down what you know and what you are looking for.

Step 3

Write down SOH CAH TOA and then circle the ratio that only includes hyp and opp.

Step 4

Write down the correct ratio, substitute the values from step 2 and work out x.

Problem solving ⚙️

Split the trapezium into a rectangle and a triangle. Remember to add on the length of the rectangle to find the length BC.

(10) Exam-style practice Grades 4–5 ✓

The diagram shows two right angled triangles, ABC and ACE.

AB = 15 cm, angle ACB = 62° and angle ECB = angle EAC = angle ABC = 90°

Work out the length CE.

You must show all your working.

[4 marks]

Trigonometry: angles

You can use trigonometry to work out unknown angles in right-angled triangles.

5 Inverse sin, cos and tan

You can use the formulae for sin, cos and tan to find an unknown angle. You will need to use the **inverse trigonometric ratios** on your calculator:

[$\sin^{-1}$] [$\cos^{-1}$] [$\tan^{-1}$]

To access these ratios, you need to press [shift] on your calculator and then one of the ratios sin, cos or tan.

Make sure your calculator is in degree mode.

5 Worked example — Grades 4–5

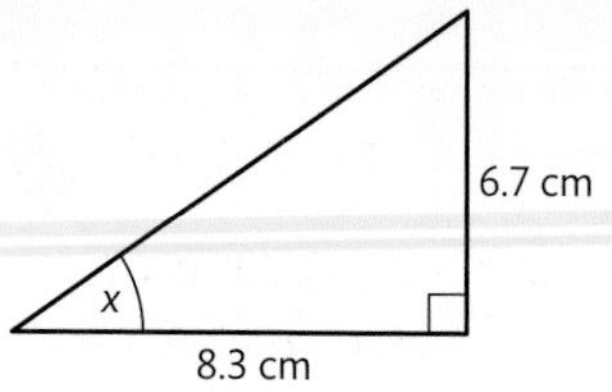
6.7 cm
x
8.3 cm

Work out the value of x.

Give your answer correct to 1 decimal place.

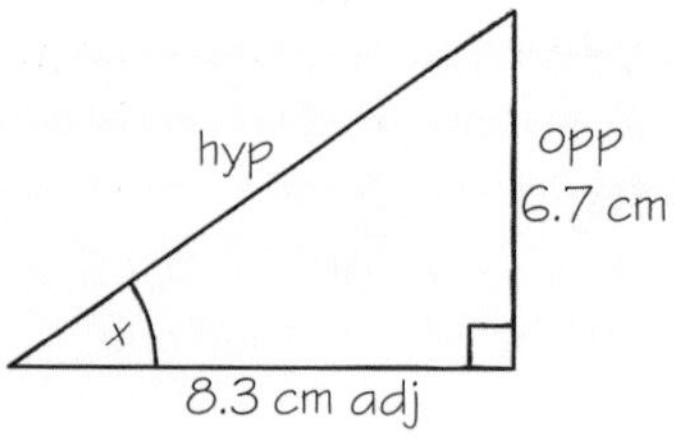
hyp
opp
6.7 cm
x
8.3 cm adj

SOH CAH [TOA]

$\tan x = \frac{\text{opp}}{\text{adj}}$ $\tan x = \frac{6.7}{8.3}$

$x = \tan^{-1}\frac{6.7}{8.3}$

$x = 38.9°$

5 Worked example — Grades 4–5

Angle $ABC = 90°$

Angle $BCD = 90°$

Work out the size of angle CDA.

Give your answer correct to 3 significant figures.

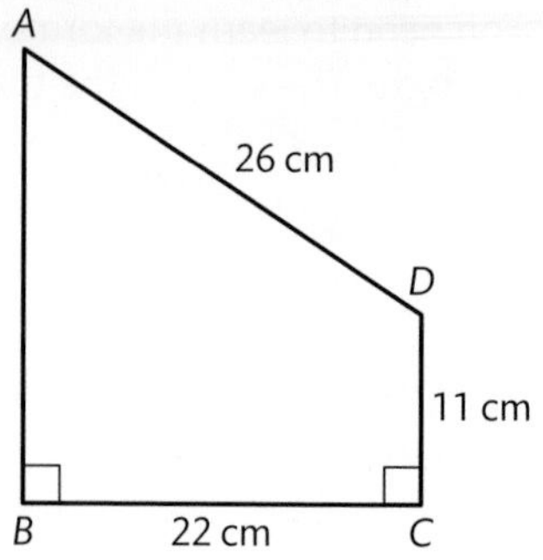
A
26 cm
D
11 cm
B
22 cm
C

SOH [CAH] TOA

$\cos x = \frac{\text{adj}}{\text{hyp}}$

$\cos x = \frac{22}{26}$

$x = \cos^{-1}\frac{22}{26}$

$x = 32.2°$

Angle $CDA = 32.2° + 90°$

$= 122.2°$

$= 122°$

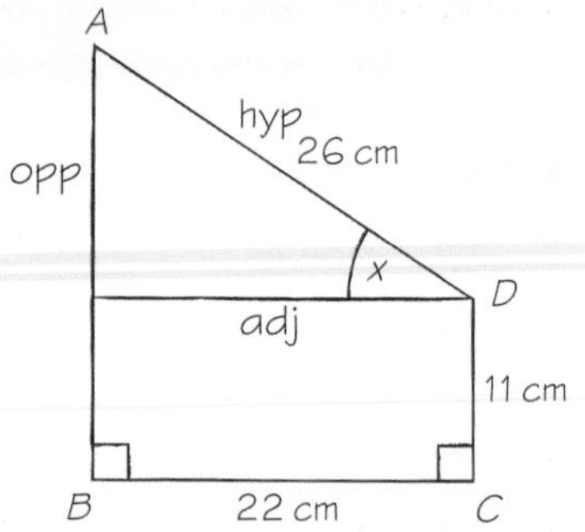
A
hyp
26 cm
opp
x
D
adj
11 cm
B
22 cm
C

Split the trapezium into a right-angled triangle and a rectangle.

Round your answer to a sensible degree of accuracy, such as 3 significant figures.

x is the angle in the right-angled triangle. You need to add 90° to find angle CDA.

On your calculator, remember to put brackets around 6.7 ÷ 8.3 or use the fraction key.

15 Exam-style practice — Grades 4–5

1 PQR is a right-angled triangle.

Angle $PQR = 90°$

$PR = 13$ cm and $QR = 5$ cm.

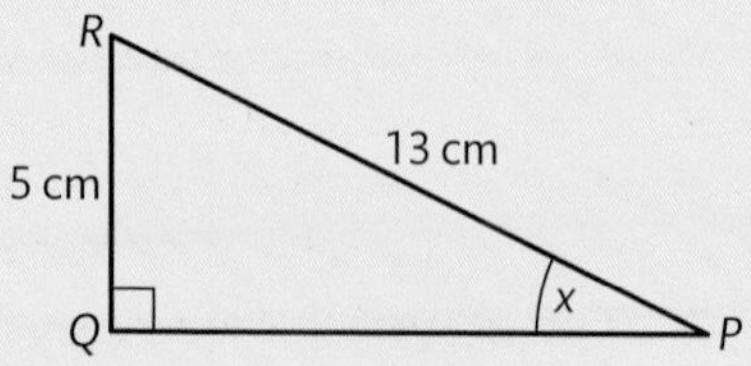
R
13 cm
5 cm
x
Q
P

Work out the value of x.

Give your answer correct to 1 decimal place.

[3 marks]

2 The diagram shows triangle ADC.

ABC is a straight line.

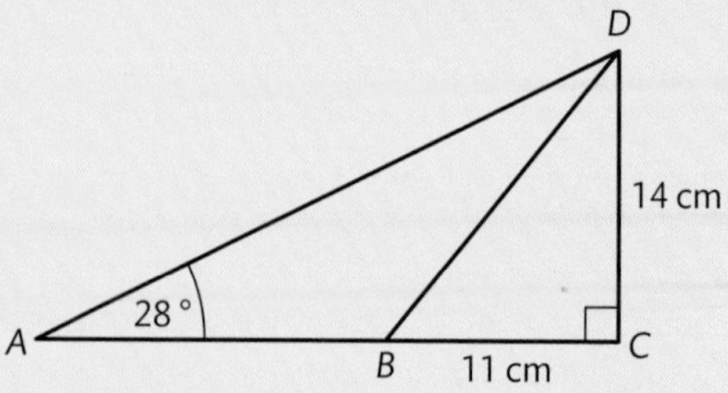
D
14 cm
28°
A
B
11 cm
C

Angle $DAC = 28°$. $DC = 14$ cm and $BC = 11$ cm.

Work out angle ADB.

Give your answer correct to 3 significant figures.

[4 marks]

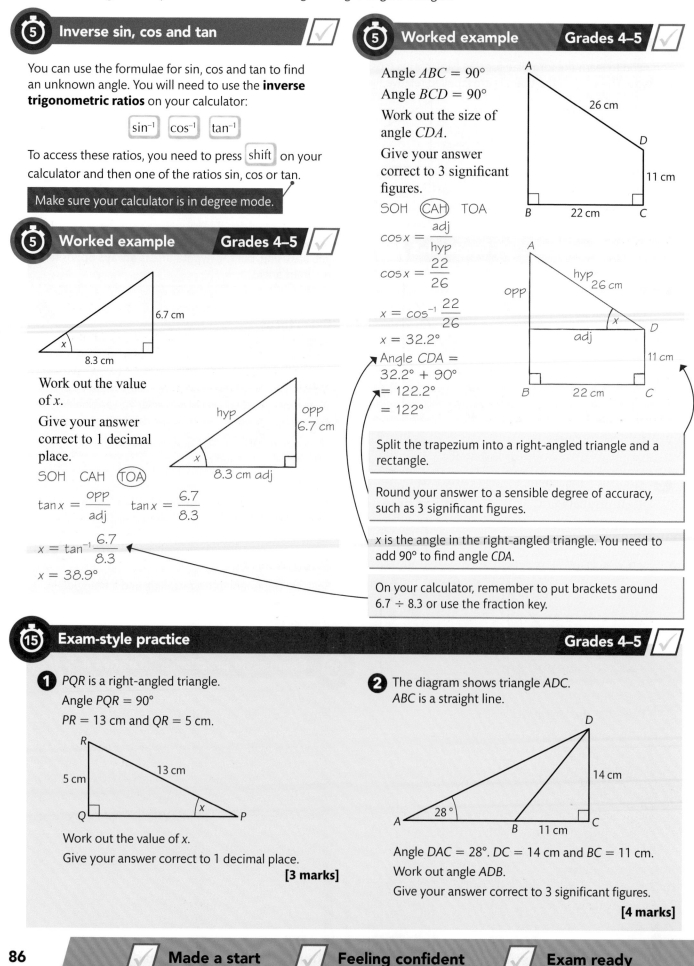

Trigonometry techniques

You need to know the values of sin, cos and tan for certain angles without a calculator. You can also use trigonometry to solve problems involving angles of elevation and depression.

5 Working without a calculator

You can use these two special triangles to find exact values for sin, cos and tan for certain angles:

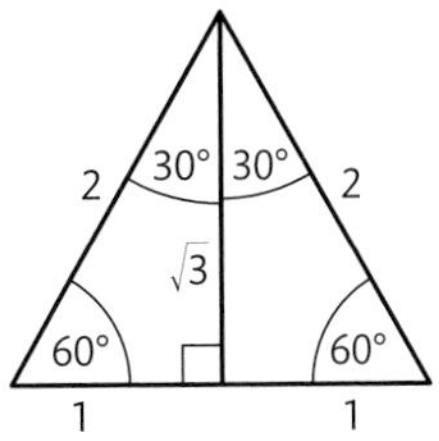

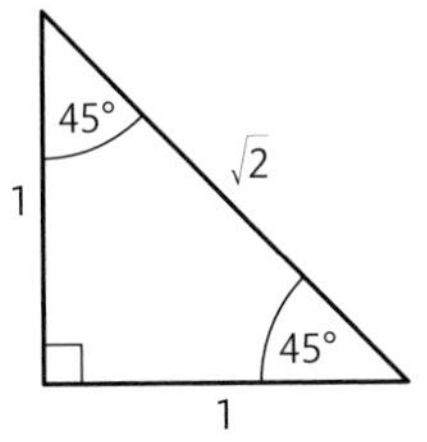

Learn these values of sin, cos and tan:

	0°	30°	45°	60°	90°
sin θ	0	$\frac{1}{2}$	$\frac{1}{\sqrt{2}}$	$\frac{\sqrt{3}}{2}$	1
cos θ	1	$\frac{\sqrt{3}}{2}$	$\frac{1}{\sqrt{2}}$	$\frac{1}{2}$	0
tan θ	0	$\frac{1}{\sqrt{3}}$	1	$\sqrt{3}$	–

2 Elevation and depression

Some angles have special names.

The **angle of elevation** is the angle between the horizontal and your line of sight for an object above you.

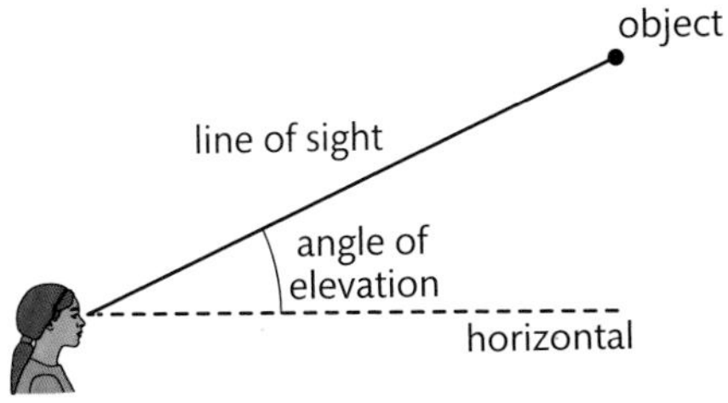

The **angle of depression** is the angle between the horizontal and your line of sight for an object below you.

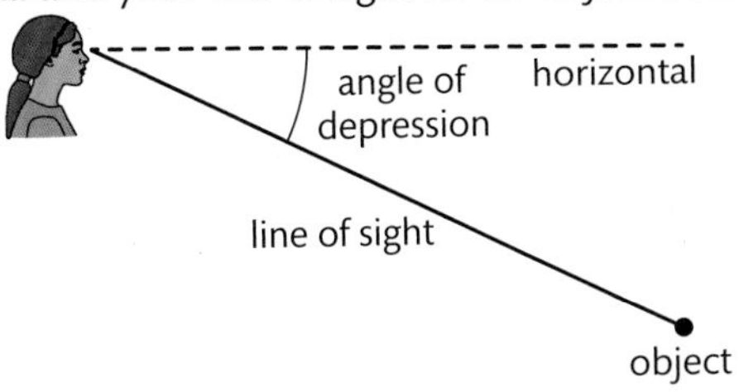

2 Checklist

- ✓ It is important that you learn values for the sin, cos and tan of 0°, 30°, 45°, 60° and 90°.
- ✓ The angle of elevation is the angle when you are looking upwards.
- ✓ The angle of depression is the angle when you are looking downwards.

2 Worked example — Grade 6

Work out the value of x.

SOH [CAH] TOA

$\cos x = \frac{15}{30} = \frac{1}{2}$

$x = 60°$

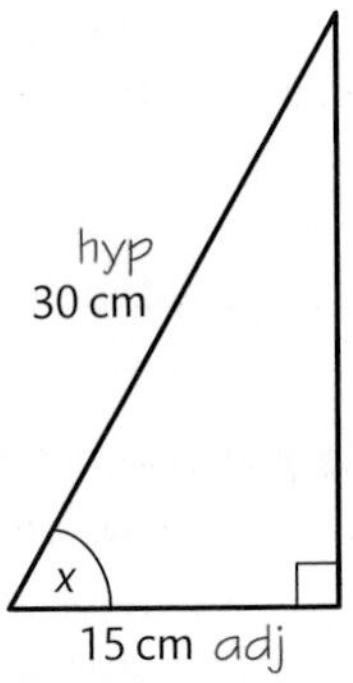

You need to be able to answer a question like this **without using a calculator**. You need to know that cos 60° = $\frac{1}{2}$.

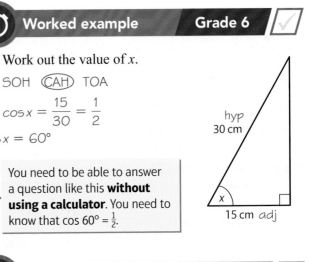

5 Worked example — Grade 6

A helicopter is hovering at a height of 1500 metres. The angle of depression from the helicopter to the landing point on the ground is 22°. Work out the distance, d, from the helicopter to the landing point.

[SOH] CAH TOA

$\sin 22° = \frac{1500}{d}$

$d \times \sin 22° = 1500$

$d = \frac{1500}{\sin 22°}$

$d = 4004$ m

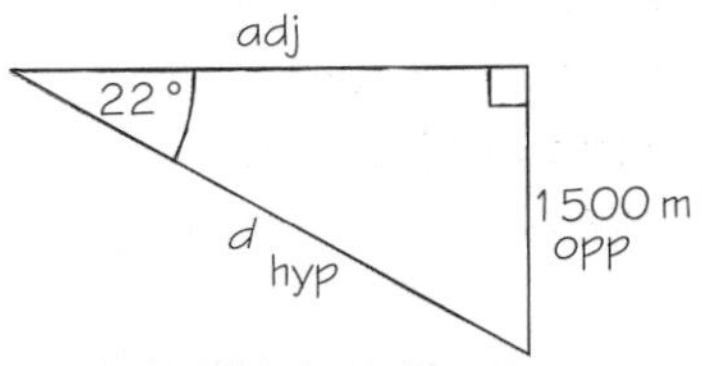

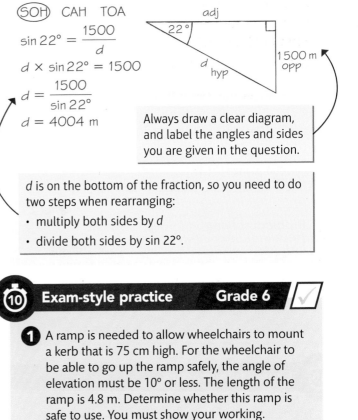

Always draw a clear diagram, and label the angles and sides you are given in the question.

d is on the bottom of the fraction, so you need to do two steps when rearranging:
- multiply both sides by d
- divide both sides by sin 22°.

10 Exam-style practice — Grade 6

1 A ramp is needed to allow wheelchairs to mount a kerb that is 75 cm high. For the wheelchair to be able to go up the ramp safely, the angle of elevation must be 10° or less. The length of the ramp is 4.8 m. Determine whether this ramp is safe to use. You must show your working.
[3 marks]

2 Work out the value of a.
Give your answer as an exact value.
[2 marks]

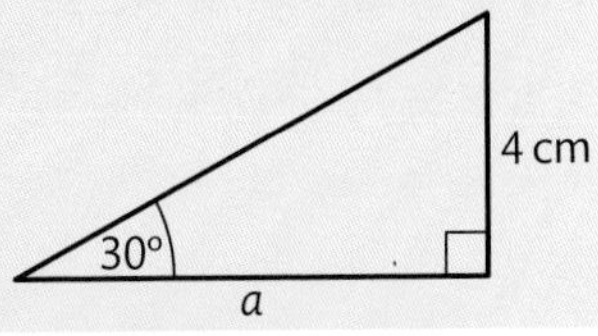

Trigonometry in 3D

Trigonometry can be applied to solve problems involving 3D shapes.

⑩ Solving 3D problems

To solve trigonometrical problems in 3D you need to break them down into two or more steps. Draw a diagram at each step. For example, this diagram shows a square-based pyramid.

How can you find the angle between a sloping edge and the base?

ABCD is a square. $AE = 6$ cm and $AB = 4$ cm. You need the size of the angle that the line *AE* makes with the plane *ABCD*. Break down this type of question and plan your strategy, **step by step**. In this example you could first find *AF* (half of *AC*), and then use that to find the angle *EAF*.

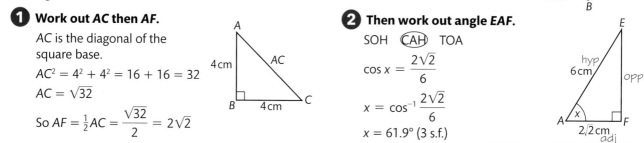

1 Work out *AC* then *AF*.

AC is the diagonal of the square base.

$AC^2 = 4^2 + 4^2 = 16 + 16 = 32$

$AC = \sqrt{32}$

So $AF = \frac{1}{2}AC = \frac{\sqrt{32}}{2} = 2\sqrt{2}$

2 Then work out angle *EAF*.

SOH ~~CAH~~ TOA

$\cos x = \dfrac{2\sqrt{2}}{6}$

$x = \cos^{-1}\dfrac{2\sqrt{2}}{6}$

$x = 61.9°$ (3 s.f.)

⑩ Worked example   Grade 6

The diagram represents a prism.

AEFD and *ABCD* are rectangles.

EB and *FC* are perpendicular to plane *ABCD*.

$AB = 50$ cm and $AD = 60$ cm.

Angle $ABE = 90°$ and angle $BAE = 30°$.

Work out the size of the angle that the line *DE* makes with the plane *ABCD*.

Give your answer correct to 1 decimal place.

Problem solving

Plan your strategy by finding lengths and/or angles step by step. For this problem you need to find the angle *EDB*. *BE* is the opposite side and *BD* is the adjacent side, so find those lengths first. Then use tan to find the angle.

Use sketches to help you see the triangles for each step.

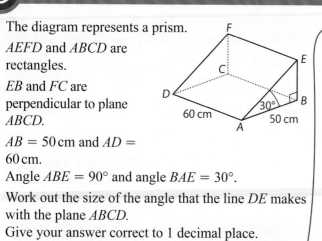

Step 1: Work out BD

$BD^2 = 60^2 + 50^2 =$
$3600 + 2500 = 6100$
$BD = \sqrt{6100}$

Step 2: Work out BE

SOH CAH ~~TOA~~

$\tan 30° = \dfrac{BE}{50}$

$BE = 50 \tan 30° = 28.8675$ cm

Step 3: Angle EDB

SOH CAH ~~TOA~~

$\tan x = \dfrac{BE}{BD} = \dfrac{28.8675}{\sqrt{6100}}$

$x = \tan^{-1}\dfrac{28.8675}{\sqrt{6100}}$

$x = 20.3°$

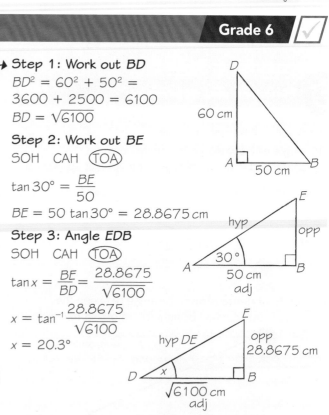

⑩ Exam-style practice   Grade 6

The diagram represents a cuboid *ABCDEFGH*.

Work out the size of the angle that the line *AG* makes with the plane *ABCD*.

Give your answer correct to 3 significant figures. **[4 marks]**

 Made a start   ✓ **Feeling confident**   ✓ **Exam ready**

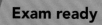

The sine rule

You can use the sine rule in any triangle to work out the length of a side or the size of an unknown angle.

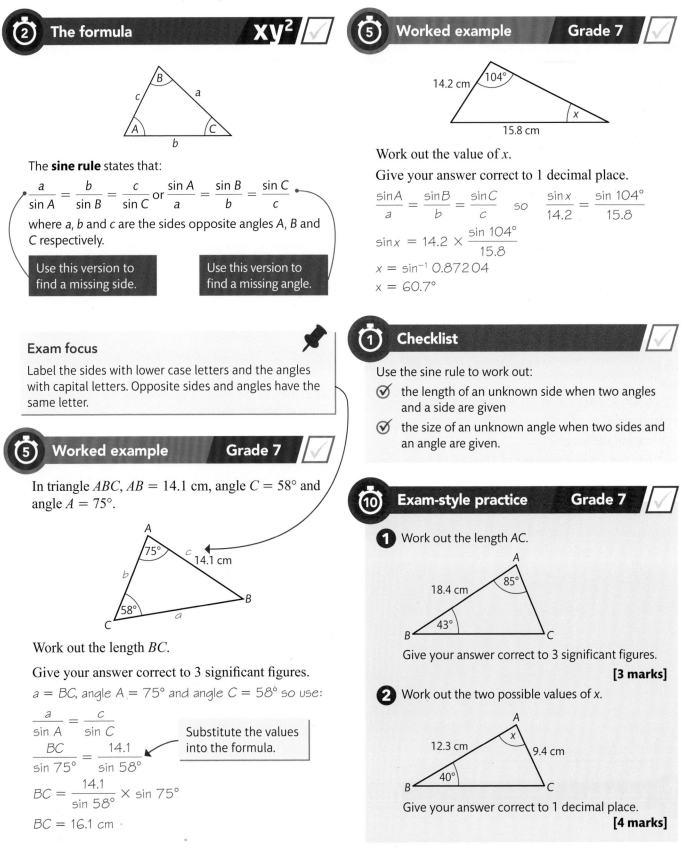

2 The formula xy^2

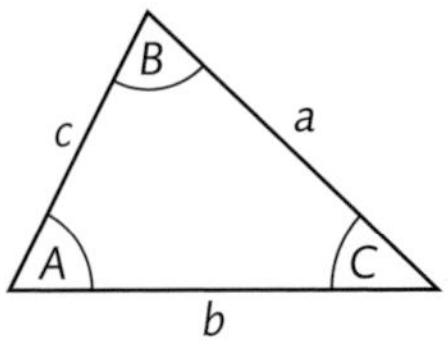

The **sine rule** states that:

$$\frac{a}{\sin A} = \frac{b}{\sin B} = \frac{c}{\sin C} \text{ or } \frac{\sin A}{a} = \frac{\sin B}{b} = \frac{\sin C}{c}$$

where *a*, *b* and *c* are the sides opposite angles *A*, *B* and *C* respectively.

Use this version to find a missing side.

Use this version to find a missing angle.

Exam focus

Label the sides with lower case letters and the angles with capital letters. Opposite sides and angles have the same letter.

5 Worked example — Grade 7

In triangle ABC, $AB = 14.1$ cm, angle $C = 58°$ and angle $A = 75°$.

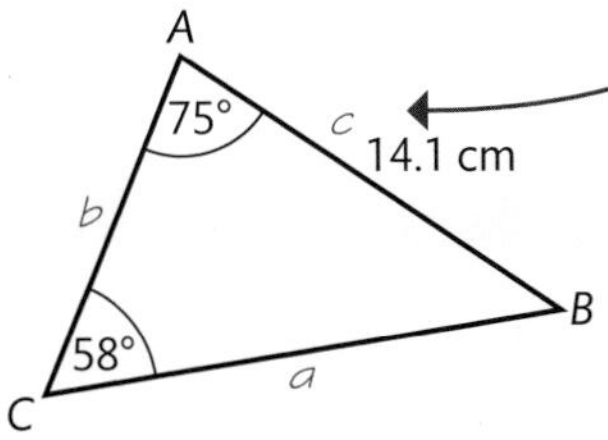

Work out the length BC.

Give your answer correct to 3 significant figures.

$a = BC$, angle $A = 75°$ and angle $C = 58°$ so use:

$$\frac{a}{\sin A} = \frac{c}{\sin C}$$

$$\frac{BC}{\sin 75°} = \frac{14.1}{\sin 58°}$$

Substitute the values into the formula.

$$BC = \frac{14.1}{\sin 58°} \times \sin 75°$$

$BC = 16.1$ cm

5 Worked example — Grade 7

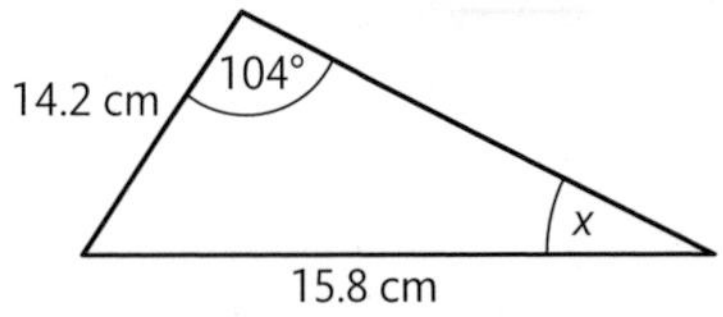

Work out the value of x.

Give your answer correct to 1 decimal place.

$$\frac{\sin A}{a} = \frac{\sin B}{b} = \frac{\sin C}{c} \text{ so } \frac{\sin x}{14.2} = \frac{\sin 104°}{15.8}$$

$$\sin x = 14.2 \times \frac{\sin 104°}{15.8}$$

$x = \sin^{-1} 0.872\,04$

$x = 60.7°$

1 Checklist

Use the sine rule to work out:

- ✓ the length of an unknown side when two angles and a side are given
- ✓ the size of an unknown angle when two sides and an angle are given.

10 Exam-style practice — Grade 7

1 Work out the length *AC*.

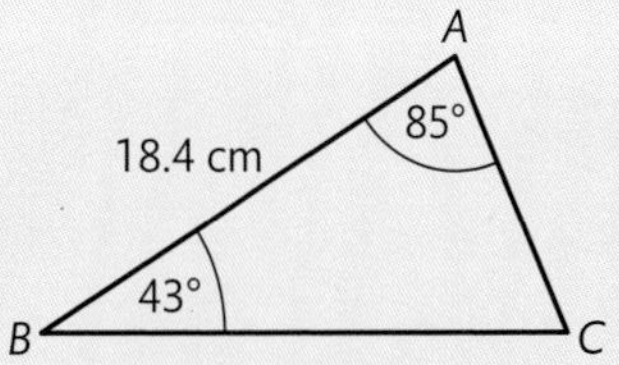

Give your answer correct to 3 significant figures.

[3 marks]

2 Work out the two possible values of *x*.

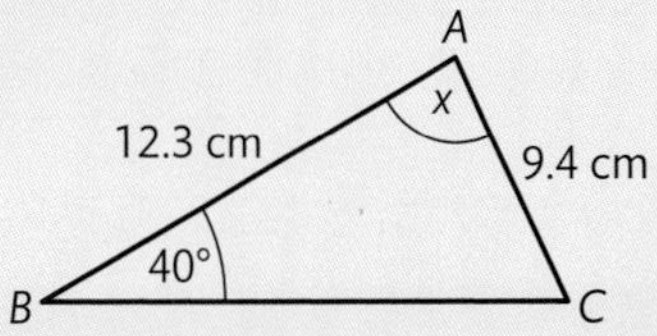

Give your answer correct to 1 decimal place.

[4 marks]

The cosine rule

The cosine rule can be used in any triangle to work out the length of a side or the size of an unknown angle.

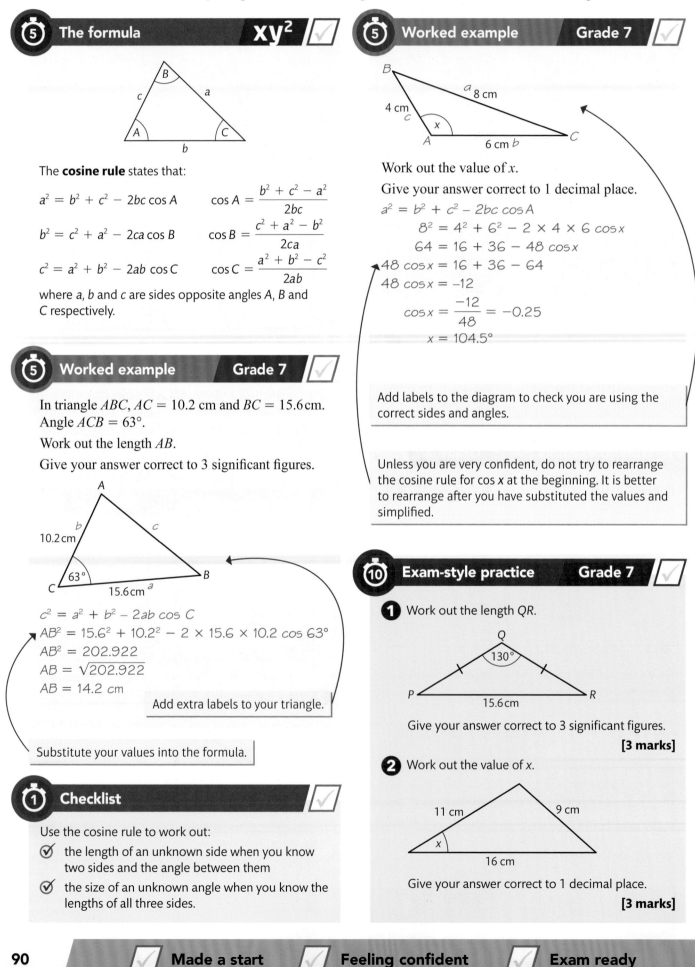

5 The formula xy^2

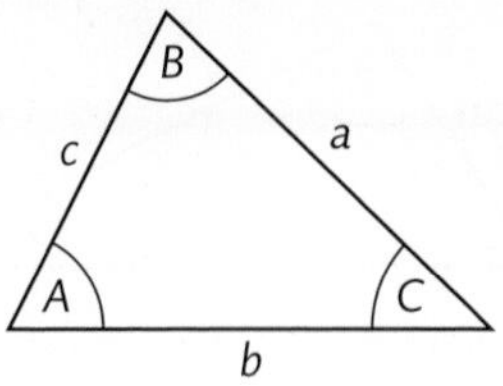

The **cosine rule** states that:

$a^2 = b^2 + c^2 - 2bc \cos A$ $\quad \cos A = \dfrac{b^2 + c^2 - a^2}{2bc}$

$b^2 = c^2 + a^2 - 2ca \cos B$ $\quad \cos B = \dfrac{c^2 + a^2 - b^2}{2ca}$

$c^2 = a^2 + b^2 - 2ab \cos C$ $\quad \cos C = \dfrac{a^2 + b^2 - c^2}{2ab}$

where *a*, *b* and *c* are sides opposite angles *A*, *B* and *C* respectively.

5 Worked example Grade 7

In triangle *ABC*, *AC* = 10.2 cm and *BC* = 15.6 cm.
Angle *ACB* = 63°.

Work out the length *AB*.

Give your answer correct to 3 significant figures.

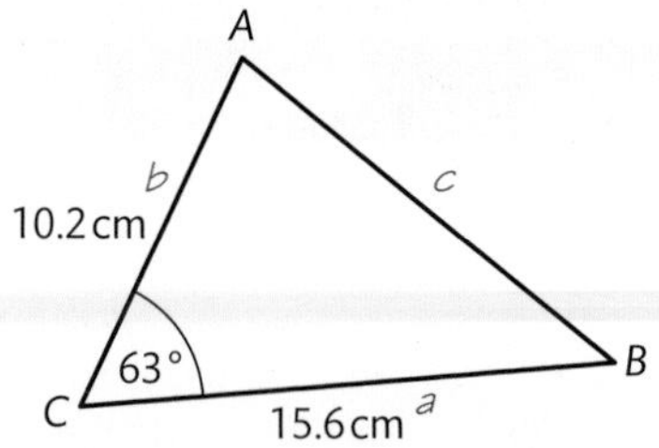

$c^2 = a^2 + b^2 - 2ab \cos C$

$AB^2 = 15.6^2 + 10.2^2 - 2 \times 15.6 \times 10.2 \cos 63°$

$AB^2 = 202.922$

$AB = \sqrt{202.922}$

$AB = 14.2$ cm

Add extra labels to your triangle.

Substitute your values into the formula.

1 Checklist

Use the cosine rule to work out:

- ✓ the length of an unknown side when you know two sides and the angle between them
- ✓ the size of an unknown angle when you know the lengths of all three sides.

5 Worked example Grade 7

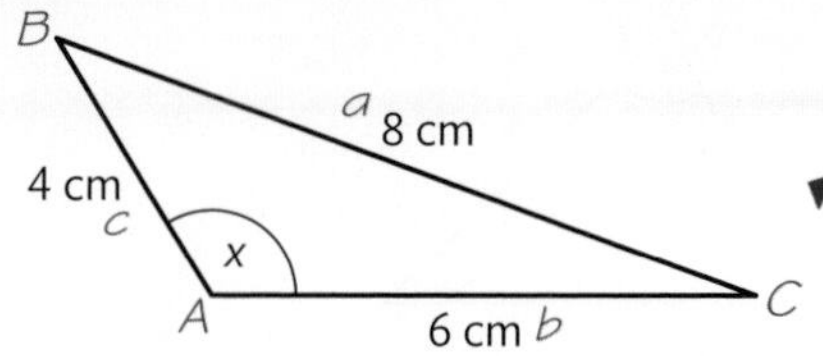

Work out the value of *x*.

Give your answer correct to 1 decimal place.

$a^2 = b^2 + c^2 - 2bc \cos A$

$8^2 = 4^2 + 6^2 - 2 \times 4 \times 6 \cos x$

$64 = 16 + 36 - 48 \cos x$

$48 \cos x = 16 + 36 - 64$

$48 \cos x = -12$

$\cos x = \dfrac{-12}{48} = -0.25$

$x = 104.5°$

Add labels to the diagram to check you are using the correct sides and angles.

Unless you are very confident, do not try to rearrange the cosine rule for cos *x* at the beginning. It is better to rearrange after you have substituted the values and simplified.

10 Exam-style practice Grade 7

1 Work out the length *QR*.

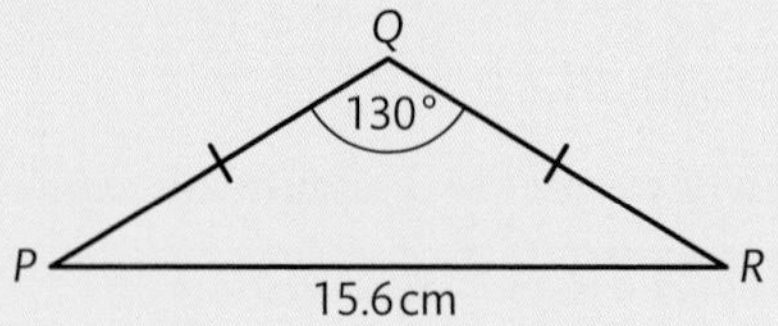

Give your answer correct to 3 significant figures.

[3 marks]

2 Work out the value of *x*.

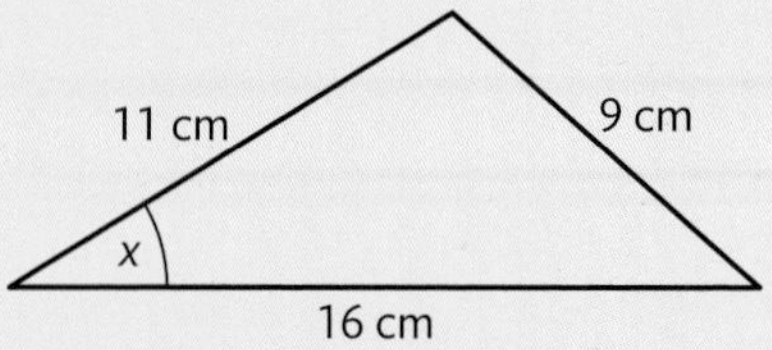

Give your answer correct to 1 decimal place.

[3 marks]

3.14

Triangles and segments

You can use a formula to work out the area of any triangle, if you know the lengths of two sides and the size of the angle between them. You will need to use this version of the formula when finding areas of segments.

(5) Area of a triangle xy^2 ✓

The area of a triangle is:
- Area $= \frac{1}{2}ab \sin C = \frac{1}{2}bc \sin A = \frac{1}{2}ca \sin B$
  where a, b and c are sides opposite angles A, B and C.

> You can only use this formula when you know two sides and the angle between them, for instance the sides a and b with the angle C.

(2) Worked example Grade 7 ✓

In triangle ABC, $AC = 10.2$ cm and $BC = 15.6$ cm and angle $ACB = 63°$.

Work out the area of triangle ABC.

Give your answer correct to 3 significant figures.

area $= \frac{1}{2}ab \sin C$
$= \frac{1}{2} \times 10.2 \times 15.6 \times \sin 63°$
$= 70.9 \text{ cm}^2$ (3 s.f.)

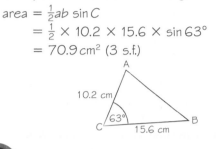
10.2 cm
63°
15.6 cm

(2) Area of a segment ✓

To find the area of a minor segment, work out the area of the sector (see page 72) and the area of the triangle and subtract.

minor segment area = sector area − triangle area

Exam focus 📌

In questions like this, leave your working in terms of π or as surds until your final answer.

(5) Worked example Grade 7 ✓

The diagram shows a sector of a circle, centre O.

The radius of the circle is 12 cm. Angle $AOC = 120°$.

Work out the area of the shaded segment ABC.
Give your answer correct to 3 significant figures.

B
A — C
12 cm   120°
O

Area of sector: $\pi r^2 \times \dfrac{\theta}{360} = \pi \times 12^2 \times \dfrac{120}{360}$
$= 48\pi$

Area of triangle: $\frac{1}{2}ab \sin C$
$= \frac{1}{2} \times 12 \times 12 \times \sin 120°$
$= 36\sqrt{3}$

Area of shaded segment $= 48\pi - 36\sqrt{3}$
$= 88.4 \text{ cm}^2$

(15) Exam-style practice Grade 7 ✓

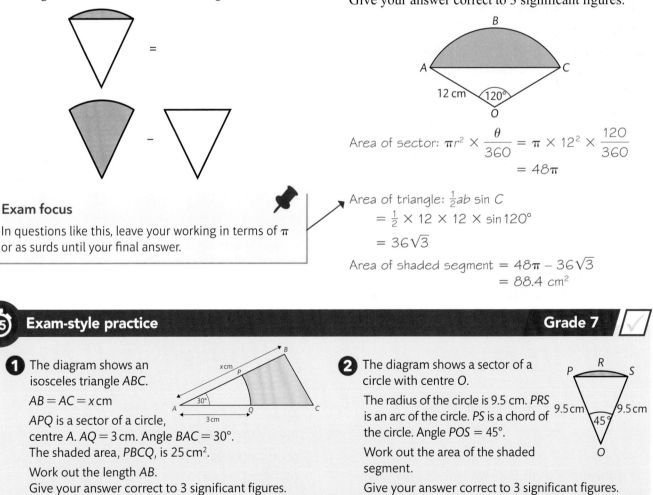

1 The diagram shows an isosceles triangle ABC.

$AB = AC = x$ cm

APQ is a sector of a circle, centre A. $AQ = 3$ cm. Angle $BAC = 30°$.
The shaded area, $PBCQ$, is 25 cm².

Work out the length AB.
Give your answer correct to 3 significant figures.

[4 marks]

2 The diagram shows a sector of a circle with centre O.

The radius of the circle is 9.5 cm. PRS is an arc of the circle. PS is a chord of the circle. Angle $POS = 45°$.

Work out the area of the shaded segment.

Give your answer correct to 3 significant figures.

[4 marks]

Vectors

Vectors are quantities, such as force or velocity, that have both **magnitude** (size) and **direction**.

⑤ About vectors ✓

A vector quantity acts in a specific direction, like the force on a snooker ball hit by a cue. In the diagram, A and B are two points. The length of the line AB represents the magnitude of the vector from A to B and the arrow shows the direction.

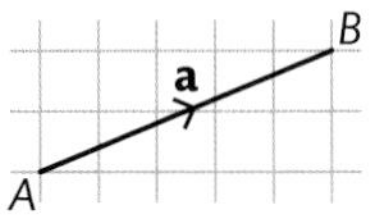

The vector from A to B can be described in three ways:

$$\overrightarrow{AB} = \mathbf{a} = \begin{pmatrix} 5 \\ 2 \end{pmatrix}$$

$\begin{pmatrix} 5 \\ 2 \end{pmatrix}$ is a **column vector** that represents a move of 5 to the right and 2 upwards. Column vectors are generally written as $\begin{pmatrix} x \\ y \end{pmatrix}$.

Vector $-\mathbf{a}$ is parallel to $\mathbf{a}$ but in the opposite direction.

⑤ Worked example — Grade 7 ✓

OAB is a triangle.
$\overrightarrow{OA} = \mathbf{a}$ and $\overrightarrow{OB} = \mathbf{b}$

> Sketch the triangle to help.

(a) Write, in terms of **a** and **b**:
(i) $\overrightarrow{AB}$

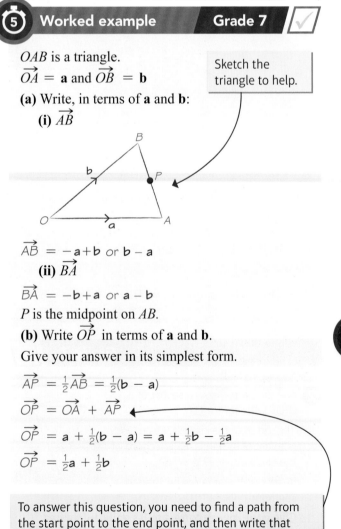

$\overrightarrow{AB} = -\mathbf{a}+\mathbf{b}$ or $\mathbf{b} - \mathbf{a}$

(ii) $\overrightarrow{BA}$

$\overrightarrow{BA} = -\mathbf{b}+\mathbf{a}$ or $\mathbf{a} - \mathbf{b}$

P is the midpoint on AB.

(b) Write $\overrightarrow{OP}$ in terms of **a** and **b**.
Give your answer in its simplest form.

$\overrightarrow{AP} = \frac{1}{2}\overrightarrow{AB} = \frac{1}{2}(\mathbf{b} - \mathbf{a})$

$\overrightarrow{OP} = \overrightarrow{OA} + \overrightarrow{AP}$

$\overrightarrow{OP} = \mathbf{a} + \frac{1}{2}(\mathbf{b} - \mathbf{a}) = \mathbf{a} + \frac{1}{2}\mathbf{b} - \frac{1}{2}\mathbf{a}$

$\overrightarrow{OP} = \frac{1}{2}\mathbf{a} + \frac{1}{2}\mathbf{b}$

> To answer this question, you need to find a path from the start point to the end point, and then write that path in terms of the vectors you are given or work out.

⑤ Parallel vectors ✓

If two vectors are parallel, one is a scalar multiple of the other.

In the diagram, all the vectors are parallel and

$\overrightarrow{AB} = \mathbf{a}$.

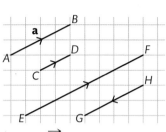

$\overrightarrow{CD}$ and $\overrightarrow{EF}$ are positive multiples of $\overrightarrow{AB}$ because they act in the same direction.

$$\overrightarrow{CD} = \begin{pmatrix} 2 \\ 1 \end{pmatrix} = \frac{1}{2}\begin{pmatrix} 4 \\ 2 \end{pmatrix} = \frac{1}{2}\mathbf{a} \quad \overrightarrow{EF} = \begin{pmatrix} 8 \\ 4 \end{pmatrix} = 2\begin{pmatrix} 4 \\ 2 \end{pmatrix} = 2\mathbf{a}$$

$$\overrightarrow{HG} = \begin{pmatrix} -4 \\ -1 \end{pmatrix} = -\mathbf{a}$$

⑤ Worked example — Grade 7 ✓

$$\mathbf{a} = \begin{pmatrix} -7 \\ 3 \end{pmatrix} \text{ and } \mathbf{b} = \begin{pmatrix} 6 \\ -1 \end{pmatrix}$$

(a) Circle the vector **a** + **b**.

$$\begin{pmatrix} 13 \\ 4 \end{pmatrix} \quad \begin{pmatrix} 1 \\ -2 \end{pmatrix} \quad \boxed{\begin{pmatrix} -1 \\ 2 \end{pmatrix}} \quad \begin{pmatrix} -13 \\ -4 \end{pmatrix}$$

$$\mathbf{a} + \mathbf{b} = \begin{pmatrix} -7 \\ 3 \end{pmatrix} + \begin{pmatrix} 6 \\ -1 \end{pmatrix} = \begin{pmatrix} -1 \\ 2 \end{pmatrix}$$

(b) Explain why $\begin{pmatrix} -6 \\ 12 \end{pmatrix}$ is parallel to **a** + **b**.

$\begin{pmatrix} -6 \\ 12 \end{pmatrix} = 6\begin{pmatrix} -1 \\ 2 \end{pmatrix}$ so $\begin{pmatrix} -6 \\ 12 \end{pmatrix}$ is a scalar multiple of **a** + **b**.

> You add or subtract column vectors by combining x-values and combining y-values.

⑩ Exam-style practice — Grade 7 ✓

ABCDEF is a regular hexagon, with centre O.
$\overrightarrow{OA} = \mathbf{a}$ and $\overrightarrow{OB} = \mathbf{b}$.

Write these vectors in terms of **a** and **b**:

(a) $\overrightarrow{EF}$ [1 mark]
(b) $\overrightarrow{AD}$ [1 mark]
(c) $\overrightarrow{AB}$ [1 mark]
(d) $\overrightarrow{CF}$ [1 mark]

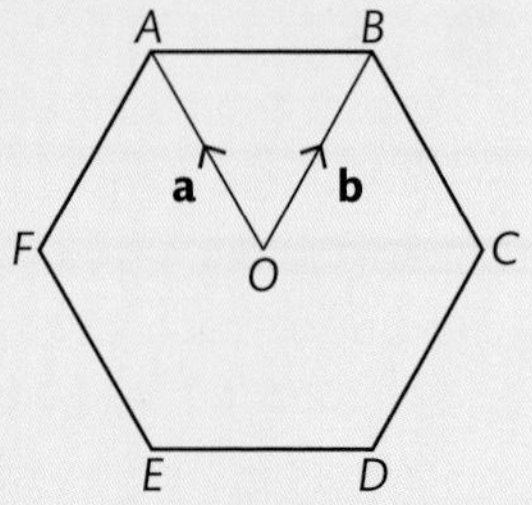

✓ **Made a start** ✓ **Feeling confident** ✓ **Exam ready**

Vector proof

You can use vector methods to prove geometric results.

⑤ Worked example **Grade 7** ✓

In parallelogram $OABC$, P is the point on AC for which $AP:PC = 2:1$.
$\overrightarrow{OA} = 3\mathbf{a}$ and $\overrightarrow{OC} = 3\mathbf{c}$.

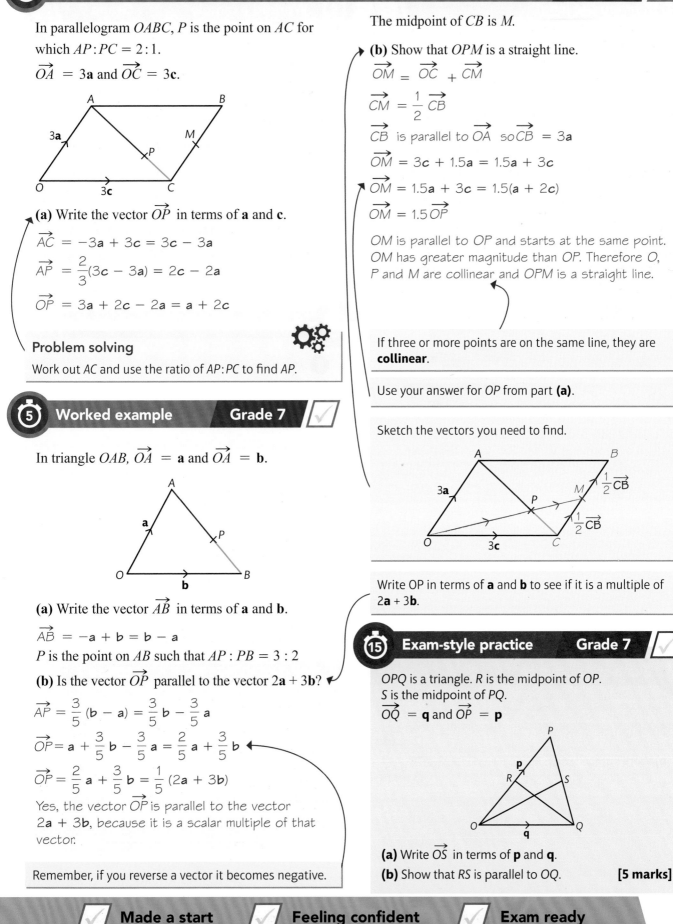

(a) Write the vector $\overrightarrow{OP}$ in terms of $\mathbf{a}$ and $\mathbf{c}$.

$\overrightarrow{AC} = -3\mathbf{a} + 3\mathbf{c} = 3\mathbf{c} - 3\mathbf{a}$

$\overrightarrow{AP} = \frac{2}{3}(3\mathbf{c} - 3\mathbf{a}) = 2\mathbf{c} - 2\mathbf{a}$

$\overrightarrow{OP} = 3\mathbf{a} + 2\mathbf{c} - 2\mathbf{a} = \mathbf{a} + 2\mathbf{c}$

Problem solving ⚙

Work out AC and use the ratio of $AP:PC$ to find AP.

⑤ Worked example **Grade 7** ✓

In triangle OAB, $\overrightarrow{OA} = \mathbf{a}$ and $\overrightarrow{OA} = \mathbf{b}$.

(a) Write the vector $\overrightarrow{AB}$ in terms of $\mathbf{a}$ and $\mathbf{b}$.

$\overrightarrow{AB} = -\mathbf{a} + \mathbf{b} = \mathbf{b} - \mathbf{a}$

P is the point on AB such that $AP:PB = 3:2$

(b) Is the vector $\overrightarrow{OP}$ parallel to the vector $2\mathbf{a} + 3\mathbf{b}$?

$\overrightarrow{AP} = \frac{3}{5}(\mathbf{b} - \mathbf{a}) = \frac{3}{5}\mathbf{b} - \frac{3}{5}\mathbf{a}$

$\overrightarrow{OP} = \mathbf{a} + \frac{3}{5}\mathbf{b} - \frac{3}{5}\mathbf{a} = \frac{2}{5}\mathbf{a} + \frac{3}{5}\mathbf{b}$

$\overrightarrow{OP} = \frac{2}{5}\mathbf{a} + \frac{3}{5}\mathbf{b} = \frac{1}{5}(2\mathbf{a} + 3\mathbf{b})$

Yes, the vector $\overrightarrow{OP}$ is parallel to the vector $2\mathbf{a} + 3\mathbf{b}$, because it is a scalar multiple of that vector.

Remember, if you reverse a vector it becomes negative.

The midpoint of CB is M.

(b) Show that OPM is a straight line.

$\overrightarrow{OM} = \overrightarrow{OC} + \overrightarrow{CM}$

$\overrightarrow{CM} = \frac{1}{2}\overrightarrow{CB}$

$\overrightarrow{CB}$ is parallel to $\overrightarrow{OA}$ so $\overrightarrow{CB} = 3\mathbf{a}$

$\overrightarrow{OM} = 3\mathbf{c} + 1.5\mathbf{a} = 1.5\mathbf{a} + 3\mathbf{c}$

$\overrightarrow{OM} = 1.5\mathbf{a} + 3\mathbf{c} = 1.5(\mathbf{a} + 2\mathbf{c})$

$\overrightarrow{OM} = 1.5\overrightarrow{OP}$

OM is parallel to OP and starts at the same point. OM has greater magnitude than OP. Therefore O, P and M are collinear and OPM is a straight line.

If three or more points are on the same line, they are **collinear**.

Use your answer for OP from part **(a)**.

Sketch the vectors you need to find.

Write OP in terms of $\mathbf{a}$ and $\mathbf{b}$ to see if it is a multiple of $2\mathbf{a} + 3\mathbf{b}$.

⑮ Exam-style practice **Grade 7** ✓

OPQ is a triangle. R is the midpoint of OP.
S is the midpoint of PQ.
$\overrightarrow{OQ} = \mathbf{q}$ and $\overrightarrow{OP} = \mathbf{p}$

(a) Write $\overrightarrow{OS}$ in terms of $\mathbf{p}$ and $\mathbf{q}$.

(b) Show that RS is parallel to OQ. **[5 marks]**

Line segments

You need to be able to find the length of a straight line on a coordinate grid. You can use Pythagoras' theorem (page 80) to do this.

5 Finding the length

You can use Pythagoras' theorem to find the length of a line segment on a coordinate grid.

Draw a right-angled triangle, using the grid lines.

Then apply Pythagoras' theorem.

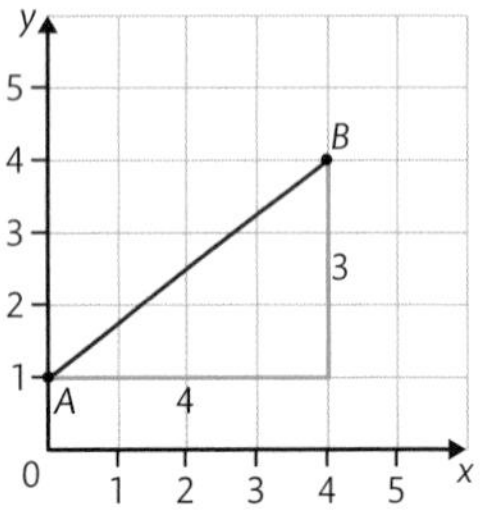

$AB^2 = 4^2 + 3^2$

$AB^2 = 16 + 9$

$AB^2 = 25$

$AB = \sqrt{25}$

$AB = 5$

You can use a formula to work out the length of any line segment, if you know the coordinates of the end points. Taking the points $A(x_1, y_1)$ and $B(x_2, y_2)$, then the length of AB is:

$$AB = \sqrt{(x_2 - x_1)^2 + (y_2 - y_1)^2}$$

You should remember this formula to find the length of a line segment.

2 The mid-point

You calculate the coordinates of the mid-point of a line segment between $A(x_1, y_1)$ and $B(x_2, y_2)$, as the mean (average) of the x values and the mean of the y values:

$$\text{mid-point of } AB = \left(\frac{x_2 + x_1}{2}, \frac{y_2 + y_1}{2}\right)$$

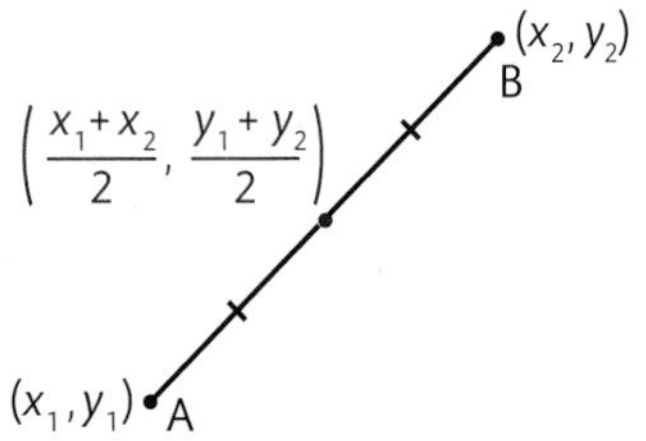

5 Worked example — Grade 7

A is the point with coordinates (3, 6).

B is the point with coordinates (9, 14).

Work out the length *AB*.

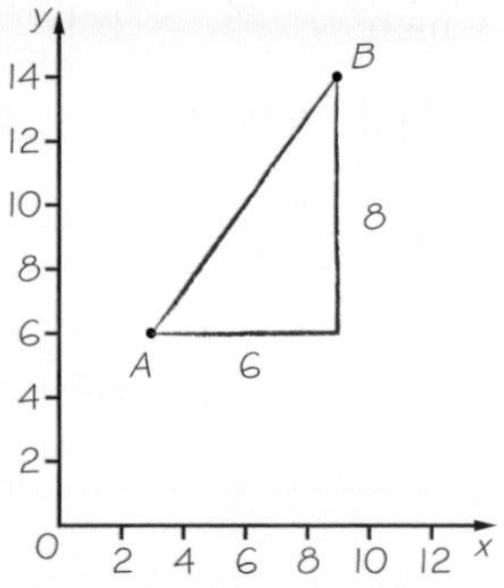

$x_2 - x_1 = 9 - 3 = 6$ and

$y_2 - y_1 = 14 - 6 = 8$

$AB^2 = 6^2 + 8^2$

$AB^2 = 100$

$AB = \sqrt{100}$

$AB = 10$

You can work out this example by drawing a right-angled triangle or by using the formula.

5 Worked example — Grade 7

1 *P* is the point with coordinates (1, 2).

Q is the point with coordinates (4, 8).

Work out the length *PQ*. Leave your answer as a simplified surd.

$PQ = \sqrt{(x_2 - x_1)^2 + (y_2 - y_1)^2}$

$PQ = \sqrt{(4 - 1)^2 + (8 - 2)^2} = \sqrt{(3)^2 + (6)^2}$

$PQ = \sqrt{9 + 36} = \sqrt{45} = \sqrt{9}\sqrt{5}$

$PQ = 3\sqrt{5}$

2 Work out the mid-point of the line between *A*(3, 7) and *B*(13, 11).

$\text{Mid-point} = \left(\frac{3 + 13}{2}, \frac{7 + 11}{2}\right) = \left(\frac{16}{2}, \frac{18}{2}\right)$

Mid-point = (8, 9)

10 Exam-style practice — Grade 7

P has coordinates (2, 3) and *Q* has coordinates (10, 7).

(a) Write the coordinates of the mid-point of the line *PQ*. **[2 marks]**

(b) Work out the length *PQ*. Leave your answer in simplified surd form. **[3 marks]**

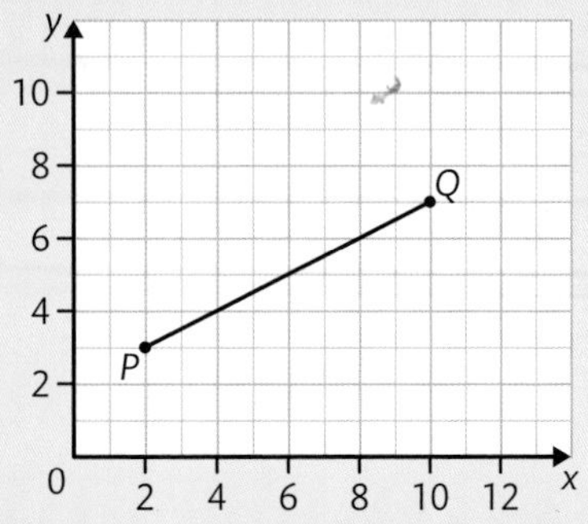

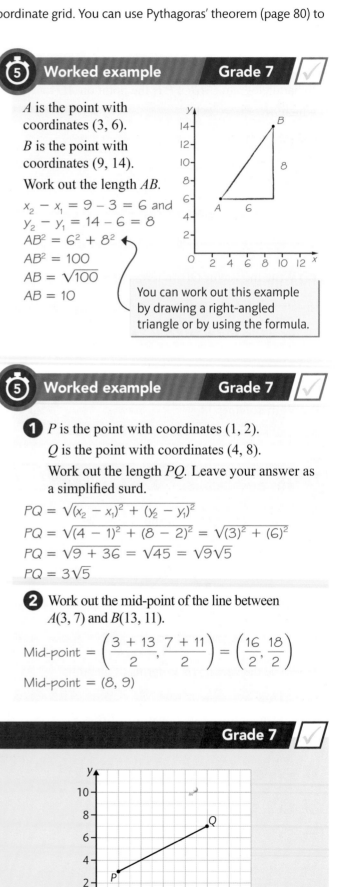

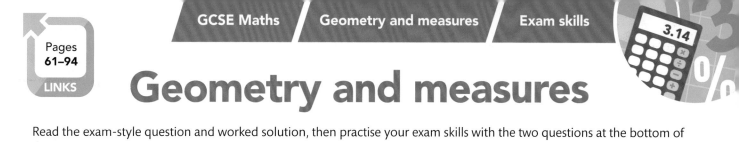

Pages
61–94
LINKS

Geometry and measures

Read the exam-style question and worked solution, then practise your exam skills with the two questions at the bottom of the page.

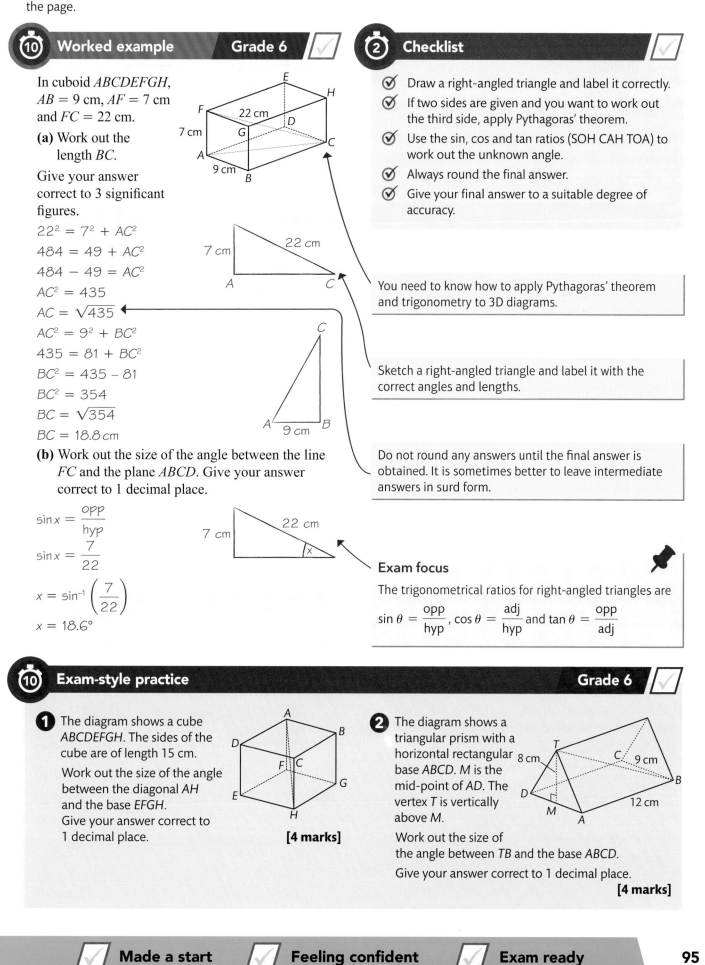

10 Worked example — Grade 6

In cuboid $ABCDEFGH$, $AB = 9$ cm, $AF = 7$ cm and $FC = 22$ cm.

(a) Work out the length BC.

Give your answer correct to 3 significant figures.

$22^2 = 7^2 + AC^2$
$484 = 49 + AC^2$
$484 - 49 = AC^2$
$AC^2 = 435$
$AC = \sqrt{435}$
$AC^2 = 9^2 + BC^2$
$435 = 81 + BC^2$
$BC^2 = 435 - 81$
$BC^2 = 354$
$BC = \sqrt{354}$
$BC = 18.8$ cm

(b) Work out the size of the angle between the line FC and the plane $ABCD$. Give your answer correct to 1 decimal place.

$\sin x = \dfrac{opp}{hyp}$
$\sin x = \dfrac{7}{22}$
$x = \sin^{-1}\left(\dfrac{7}{22}\right)$
$x = 18.6°$

2 Checklist

☑ Draw a right-angled triangle and label it correctly.
☑ If two sides are given and you want to work out the third side, apply Pythagoras' theorem.
☑ Use the sin, cos and tan ratios (SOH CAH TOA) to work out the unknown angle.
☑ Always round the final answer.
☑ Give your final answer to a suitable degree of accuracy.

You need to know how to apply Pythagoras' theorem and trigonometry to 3D diagrams.

Sketch a right-angled triangle and label it with the correct angles and lengths.

Do not round any answers until the final answer is obtained. It is sometimes better to leave intermediate answers in surd form.

Exam focus

The trigonometrical ratios for right-angled triangles are

$$\sin \theta = \frac{opp}{hyp}, \cos \theta = \frac{adj}{hyp} \text{ and } \tan \theta = \frac{opp}{adj}$$

10 Exam-style practice — Grade 6

1 The diagram shows a cube $ABCDEFGH$. The sides of the cube are of length 15 cm.

Work out the size of the angle between the diagonal AH and the base $EFGH$.

Give your answer correct to 1 decimal place.

[4 marks]

2 The diagram shows a triangular prism with a horizontal rectangular base $ABCD$. M is the mid-point of AD. The vertex T is vertically above M.

Work out the size of the angle between TB and the base $ABCD$.

Give your answer correct to 1 decimal place.

[4 marks]

☑ **Made a start**   ☑ **Feeling confident**   ☑ **Exam ready**

Probability

Probability is the likelihood of an event occurring.

(2) Equally likely outcomes

For **equally likely** outcomes the probability that a particular outcome occurs is

$$\text{probability} = \frac{\text{number of successful outcomes}}{\text{total number of outcomes}}$$

The sum of the probabilities of all the possible outcomes is **1**.

P(outcome occurs) = 1 − P(outcome does not occur)

(2) Types of event

Independent events can happen at the same time but the outcome of one does not affect the outcome of the other. For independent events A and B:

P(A and B) = P(A) × P(B)

Mutually exclusive events cannot happen at the same time. For mutually exclusive events A and B:

P(A or B) = P(A) + P(B)

(10) Worked example — Grade 5

1 A box contains red, green, blue and orange pencils. Burt selects a pencil from the box at random. The probability that he selects each colour is shown in the table.

Colour	red	green	blue	orange
Probability	0.24	0.18		

The probability that Burt selects a green pencil is twice the probability that he selects a blue pencil.

Work out the probability that he will:

(a) take a blue pencil

P(blue) = 0.18 ÷ 2 = 0.09

(b) not take a blue pencil ← P(not B) = 1 − P(B)

P(not blue) = 1 − 0.09 = 0.91

(c) take a red pencil or a green pencil ←

P(red or green) = 0.24 + 0.18 = 0.42

(d) take an orange pencil.

P(orange) = 1 − (0.24 + 0.18 + 0.09)
= 1 − 0.51 = 0.49

2 Ravina rolls a fair dice and tosses a fair coin. Work out the probability that she gets:

(a) a six and a head

$$\frac{1}{6} \times \frac{1}{2} = \frac{1}{12}$$

(b) an even number and a tail.

$$\frac{1}{2} \times \frac{1}{2} = \frac{1}{4}$$

Problem solving

Burt cannot choose a red pencil and a green pencil at the same time so these events are mutually exclusive.
P(R or G) = P(R) + P(G)

(5) Worked example — Grade 5

There are 35 boys and 48 girls in a club. The ratio of the number of boys who play pool to the number of boys who do **not** play pool is 2:5. The ratio of the number of girls who play pool to the number of girls who do **not** play pool is 1:3. The club leader picks at random one child from all of those who play pool.

Work out the probability that this child is a girl.

The number of boys who play pool
$= \frac{2}{7} \times 35 = 10$

The number of girls who play pool
$= \frac{1}{4} \times 48 = 12$

Total number who play pool = 10 + 12 = 22

The probability that this child is a girl $= \frac{12}{22}$

(10) Exam-style practice — Grade 5

1 A bag contains only red counters, yellow counters and blue counters. Ken takes at random a counter from the bag. The table shows each of the probabilities.

Colour	red	yellow	blue
Probability	0.4	0.5	

There are 20 red counters. How many blue counters are there? **[3 marks]**

2 Sandeep has a biased coin. The probability that the coin will land tails up is 0.7. Sandeep is going to throw the coin 3 times. He says the probability that the coin will land heads up 3 times is less than 0.1.

Show whether Sandeep is correct. You must show all your working. **[3 marks]**

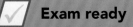 Made a start | Feeling confident | Exam ready

Relative frequency

Relative frequency is how often an outcome occurs, in a series of trials. The more times a trial is carried out, the closer a relative frequency gets to the theoretical frequency. Relative frequency can be used as an estimate for the probability of the outcome.

$$\text{Relative frequency} = \frac{\text{frequency of outcome}}{\text{total frequency}}$$

10 Worked example — Grade 5

Rebecca carries out a survey about the number of times customers go to a supermarket. She asks, at random, 80 customers how many times they went to the supermarket last month. The table shows Rebecca's results.

Visits	0	1	2	3	4	5 or more
Frequency	6	24	17	16	10	7

A customer is chosen at random. Estimate the probability that the customer went to the supermarket:

(a) exactly 2 times

$$P(\text{exactly 2 times}) = \frac{17}{80}$$

(b) 3 or more times.

$$P(\text{3 or more times}) = \frac{16 + 10 + 7}{80} = \frac{33}{80}$$

From the table, total frequency is 6 + 24 + 17 + 16 + 10 + 7 = 80.

5 Worked example — Grade 5

Here is some information about the total rainfall, in millimetres, recorded at 100 weather stations over one month.

Rainfall, x (mm)	Frequency
$0 < x \leqslant 15$	18
$15 < x \leqslant 30$	29
$30 < x \leqslant 50$	32
$50 < x \leqslant 80$	21

Another weather station in the same area is chosen.

(a) Estimate the probability that the rainfall recorded by the new station was:

   (i) between 15 mm and 30 mm

$$P(\text{between 15 mm and 30 mm}) = \frac{29}{100} = 0.29$$

   (ii) over 30 mm.

$$P(\text{over 30 mm}) = \frac{32 + 21}{100} = \frac{53}{100} = 0.53$$

(b) Comment on the accuracy of your estimates.

The estimates are accurate because the sample size is large.

10 Exam-style practice — Grade 5

1 A bag contains red, green, yellow and blue counters. The table gives information about the counters.

Colour	red	green	yellow	blue
Frequency	28	21	24	17

A counter is to be taken at random from the bag. Work out the probability that the counter will be:

(a) yellow   **[1 mark]**

(b) red or green or blue   **[2 marks]**

(c) not green.   **[2 marks]**

2 The table shows some information about the time, in seconds, people spent in a queue at a supermarket.

Time spent x (seconds)	Frequency
$0 < x \leqslant 50$	3
$50 < x \leqslant 100$	4
$100 < x \leqslant 150$	6
$150 < x \leqslant 200$	2

A new customer visits the supermarket.

(a) Estimate the probability that the customer spends:

   (i) more than 150 seconds in a queue   **[1 mark]**

   (ii) less than 100 seconds in a queue.   **[2 marks]**

(b) Comment on the accuracy of your estimates.   **[1 mark]**

Venn diagrams

A Venn diagram represents connections between different sets of data. For a diagram showing two sets, the overlapping region of the circles represents the elements that are in both sets. Any values outside of the circles are not in either set.

⑤ What is a Venn diagram?

A **Venn diagram** shows information about sets or groups of data. Consider two events A and B.

The probability of event A is written as P(A). The probability that A does not happen is $P(A') = 1 - P(A)$, and is displayed outside of the circle.

The intersection between the circles shows the elements that are in both sets, $A \cap B$.

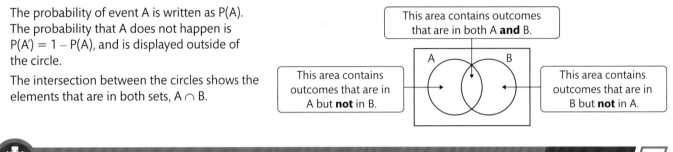

This area contains outcomes that are in both A **and** B.

This area contains outcomes that are in A but **not** in B.

This area contains outcomes that are in B but **not** in A.

⑤ Worked example   Grade 6

1 In a group of 30 students, 13 play football, 19 play hockey, 7 play football and hockey.

(a) Represent this information on a Venn diagram.

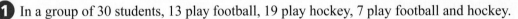

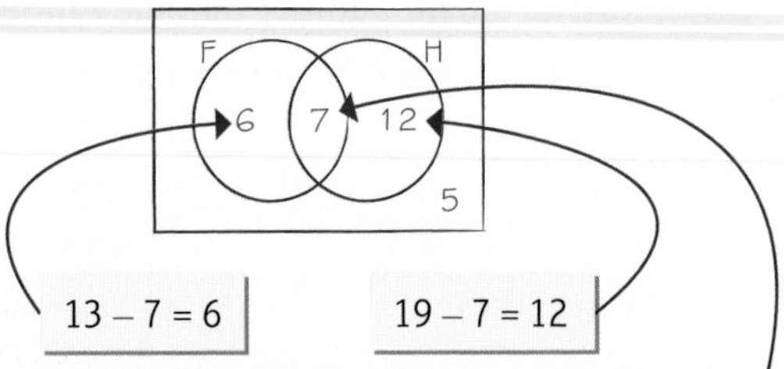

F   H
6   7   12
5

13 − 7 = 6   19 − 7 = 12

Start from the intersection and work your way outwards. Write 7, then work out how many play football but not hockey, and how many play hockey but not football.

Exam focus

You do not need to simplify fractions in probability questions in your exam.

b) Work out the probability that a student chosen at random plays football only.

$$P(\text{football only}) = \frac{6}{30}$$

(c) Work out the probability that a student plays hockey only.

$$P(\text{hockey only}) = \frac{12}{30}$$

(d) Work out the probability that a student plays football or hockey but not both.

$$P(\text{football or hockey but not both}) = \frac{6 + 12}{30} = \frac{18}{30}$$

(e) Work out the probability that a student does not play football or hockey.

$$P(\text{does not play football or hockey}) = \frac{5}{30}$$

⑮ Exam-style practice   Grade 6

1 Simon gathered some information about the pet dogs and pet cats around the estate he lives in. There are 90 families in this estate. 65 families have a dog, 34 have a cat and 12 have a dog and a cat. No family had more than 1 cat or more than 1 dog.

   (a) Draw a Venn diagram to represent this information.   **[3 marks]**

   (b) What is the probability that a family, chosen at random, has neither a dog nor a cat?   **[2 marks]**

2 A running club has 60 members. 44 of the members take part in a cross country run, 27 of the members take part in a marathon and 6 of the members do not run in the cross country or the marathon.

   Work out the probability that a member only takes part in the cross country run or in the marathon but not both.
                                                                                [4 marks]

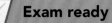 Made a start   Feeling confident   Exam ready

Conditional probability

Conditional probability is the probability of an event occurring, given that another event has occurred.

 Calculating conditional probablity ✓

You can work out conditional probability from Venn diagrams and from two-way tables.

It is important to remember that the probability of the second event can be influenced by the outcome of the first event.

P(A|B) means the probability of A occurring, given that B has already occurred.

> It is a good idea to write out the totals for each row and column.

 Worked example **Grade 9** ✓

The Venn diagram gives information about the different sports, netball and hockey, played by a group of students.

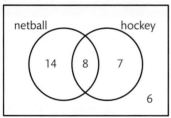

netball | hockey

14 | 8 | 7

6

A student is chosen at random.

(a) Work out the probability that the student plays netball, given that the student also plays hockey.

15 students play hockey.

$$P(\text{netball}|\text{hockey}) = \frac{8}{8 + 7} = \frac{8}{15}$$

(b) Work out the probability that the student plays both sports, given that the student plays at least one sport.

29 students play at least one sport.

P(both sports|at least one sport)

$$= \frac{8}{14 + 8 + 7} = \frac{8}{29}$$

> The number of students who play both sports is 8. Make sure you show all your working.

 Worked example **Grade 9** ✓

The two-way table gives information about the numbers of teachers who work in state and independent schools who attended a conference.

	State	Independent	
Male	59	78	137
Female	62	51	113
	121	129	

(a) One of the male teachers is chosen at random. What is the probability that he works in an independent school?

$$P(\text{independent}|\text{male}) = \frac{78}{59 + 78} = \frac{78}{137}$$

(b) One of the teachers who works in a state school is chosen at random. What is the probability that this teacher is female?

$$P(\text{female}|\text{state}) = \frac{62}{59 + 62} = \frac{62}{121}$$

 **Exam-style practice** **Grade 9** ✓

1 80 children were asked what type of drink they took to school. 38 said they took water, 27 took squash and 12 took both water and squash.

(a) Show this information on a Venn diagram. **[3 marks]**

(b) Work out the probability that a child takes squash, given that they take water. **[2 marks]**

(c) Given that a child takes squash, find the probability that this child also takes water. **[2 marks]**

2 The two-way table gives information about the weight of some adults.

	Underweight	Just right	Overweight
Male	15	18	22
Female	19	28	35

(a) One of the males is picked at random. Work out the probability that he is overweight. **[2 marks]**

(b) One of the adults who is underweight is chosen. Work out the probability that this is a female. **[2 marks]**

Tree diagrams

A **tree diagram** represents a sequence of two or more events. It can be useful in solving probability problems.

10 Using tree diagrams

Tree diagrams record all possible outcomes in a clear and logical manner. The probabilities are written on the branches.

As a general rule:

Multiply along the branches and add between the branches.

Tree diagrams can be used to solve problems involving conditional probability (page 97).

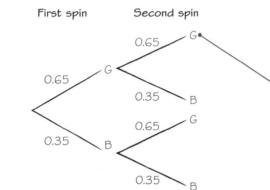

In a game where a spinner can land on green or on blue, the probability that the spinner will land on green is 0.65. Tree diagrams have probabilities written on the branches, so multiplying along the branches is the event that the spinner lands on green twice. It happens with probability $0.65 \times 0.65 = 0.4225$.

5 Worked example · Grades 6–7

There are 12 pens in a box.

Five of the pens are yellow.

Seven of the pens are red.

Andy takes at random a pen from the box. He puts the pen in his pocket. He then takes at random another pen from the box.

(a) Show this information on a tree diagram.

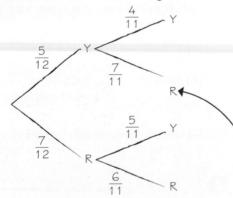

(b) Work out the probability that Andy takes two pens of the same colour.

$P(\text{both the same colour}) = YY \text{ or } RR$

$= \left(\dfrac{5}{12} \times \dfrac{4}{11}\right) + \left(\dfrac{7}{12} \times \dfrac{6}{11}\right)$

$= \dfrac{20}{132} + \dfrac{42}{132}$

$= \dfrac{62}{132}$

This type of question is **without replacement**. The number of possible outcomes is reduced by 1 each time. This means that the denominator of the second probability is reduced by 1.

2 Checklist

- ☑ Along the branches you multiply the probabilities.
- ☑ The sum of all the probabilities at the ends of the branches is 1.
- ☑ Between the branches you add the probabilities.
- ☑ If an object is put back then this is known as **with replacement**.
- ☑ If an object is not put back then the probability is known as **without replacement**.

15 Exam-style practice · Grades 6–7

1 The probability that it will rain on Saturday is 0.7.

When it rains on Saturday, the probability that it will rain on Sunday is 0.75.

When it does **not** rain on Saturday, the probability that it will rain on Sunday is 0.4.

(a) Show this information on a tree diagram. **[3 marks]**

(b) Work out the probability that it will rain on both Saturday and Sunday. **[2 marks]**

(c) Work out the probability that it will rain on at least one of the two days. **[2 marks]**

2 Amina has 20 medals in a box. She has 11 gold medals, 6 silver medals and 3 bronze medals.

She takes at random two medals from the box. Work out the probability that the two medals are **not** the same type. **[5 marks]**

✓ **Made a start**   ✓ **Feeling confident**   ✓ **Exam ready**

Pages
96–100

LINKS

Probability

Read the exam-style question and worked solution, then practise your exam skills with the two questions at the bottom of the page.

⑩ Worked example — Grades 6–7 ✓

There are 15 counters in a bag. Of these 9 are green and 6 are red.

Rishi takes at random a counter from the bag. He puts the counter back in the bag. He then takes at random another counter from the bag.

(a) Work out the probability that Rishi takes one counter of each colour. You must show your working.

[3 marks]

$P(1 \text{ each colour}) = P(R \text{ and } G) + P(G \text{ and } R)$

$= \left(\dfrac{6}{15} \times \dfrac{9}{15}\right) + \left(\dfrac{9}{15} \times \dfrac{6}{15}\right)$

$= \dfrac{54}{225} + \dfrac{54}{225}$

$= \dfrac{108}{225}$

Maria takes at random a counter from the bag. She keeps the counter. She then takes at random another counter from the bag.

(b) Work out the probability that Maria takes one counter of each colour. You must show your working.

[4 marks]

$P(1 \text{ each colour}) = P(R \text{ and } G) + P(G \text{ and } R)$

$= \left(\dfrac{6}{15} \times \dfrac{9}{14}\right) + \left(\dfrac{9}{15} \times \dfrac{6}{14}\right)$

$= \dfrac{54}{210} + \dfrac{54}{210}$

$= \dfrac{108}{210}$

Problem solving

Sometimes it is helpful to draw a tree diagram even if one is not asked for.

② Replacement checklist ✓

- ☑ With replacement the probabilities remain the same.
- ☑ Without replacement the probabilities change.
- ☑ AND means multiplication of probabilities.
- ☑ OR means addition of probabilities.

You need to know that Rishi could choose a red counter followed by a green counter **or** a green counter followed by a red counter, so there are two possible ways of achieving this outcome.

Part **(a)** is selection **with replacement**, so the number of objects does not change, therefore the value of the denominator does not change.

Revise how to manipulate fractions on page 1.

In part **(b)**, Maria does not put the counter back in the bag. It is an example of selection **without replacement**. The number of objects changes and so the value of the denominator is reduced by 1.

⑩ Exam-style practice — Grades 6–7 ✓

1 Harold has nine socks. Five are black, two are white and two are grey.

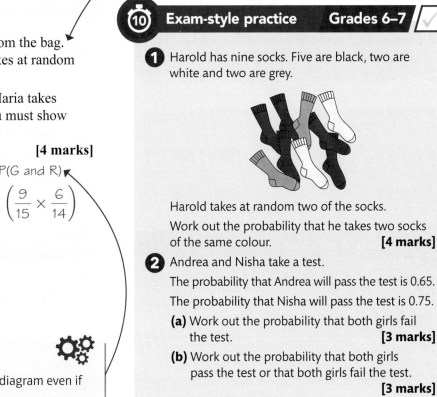

Harold takes at random two of the socks.

Work out the probability that he takes two socks of the same colour. **[4 marks]**

2 Andrea and Nisha take a test.

The probability that Andrea will pass the test is 0.65.

The probability that Nisha will pass the test is 0.75.

(a) Work out the probability that both girls fail the test. **[3 marks]**

(b) Work out the probability that both girls pass the test or that both girls fail the test. **[3 marks]**

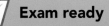

Sampling

The way in which data is collected can have a big impact on the validity of any conclusions.

⑤ What is data?

Discrete data can only take certain values, for example, hair colour, shoe size.
Continuous data can take any value in an interval, for example, length, area.
Primary data is data you collect yourself, for example, by conducting a survey.
Secondary data is data that someone else, such as a research organisation, has collected.

⑤ About sampling

A **population** is every single person or item you might be interested in.

A **sample** is a selection or people or items taken from the whole population.

To represent the population accurately, the sample should be **random** and free from **bias**.

Bias can occur in many ways. For example:

- The sample may not be representative of the population, for example, a weighted dice that purposefully gives one result more often the others.
- The population may not be appropriate for the survey being carried out, for example, it may leave out a group who ought to be included.

How do you avoid bias?
- Use a larger sample.
- Use a random sample.

Advantages and disadvantages of samples

👍 Cheaper than using the entire population
👍 Less time-consuming
👍 Less data to handle (easier)
👎 Not completely representative
👎 May be biased

> Try to give answers in the context of the question not just quote standard answers.

⑤ Worked example — Grade 5

Ellen asks 10 of her friends at her running club to take part in a survey. Give **one** reason why this may not be a suitable sample.

The sample could be biased as it only includes her friends who go to the running club.

You could also suggest that the sample size is too small or may not be representative of all the members of the club.

⑤ Worked example — Grade 5

❶ Ray wants to find out how much time people spend watching football on television. He is going to use a questionnaire to carry out a survey. Ray asks the boys in his class to complete his questionnaire. Give **two** reasons why his sample is biased.

He only asks boys, and they are all in the same class/year/school.

⑤ Exam-style practice — Grade 5

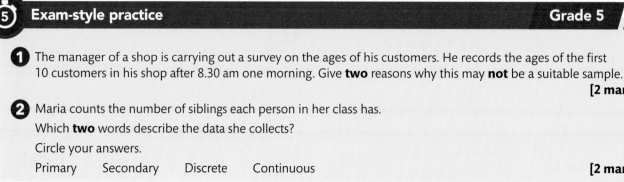

❶ The manager of a shop is carrying out a survey on the ages of his customers. He records the ages of the first 10 customers in his shop after 8.30 am one morning. Give **two** reasons why this may **not** be a suitable sample.

[2 marks]

❷ Maria counts the number of siblings each person in her class has.
Which **two** words describe the data she collects?
Circle your answers.

Primary   Secondary   Discrete   Continuous

[2 marks]

☑ **Made a start**   ☑ **Feeling confident**   ☑ **Exam ready**

BBC

3.14

Mean, median and mode

An average is a value that represents the data set. Three different types of average are used to analyse and compare data. An **outlier** is a value that does not fit with the trend.

(5) **The three averages**

How to find the average	Advantages and disadvantages
Mean Add up all the values and then divide the total by the number of values.	👍 Uses all the data 👎 Affected by extreme values (outliers)
Mode Identify the value that occurs the most – that has the highest frequency.	👍 Can apply to all kinds of data, including non-numerical, such as colours or makes of car 👎 May not exist if no value is repeated, or there may be two or more modes
Median Arrange the values numerically and find the middle value. For an even number of values, take the mean of the middle two.	👍 Not affected by extreme values 👎 The value may not exist

(10) **Worked example** Grade 5

1 Nisha has five cards. She wants to write down a number on each card such that:

> the mode of the five numbers is 9
> the median of the five numbers is 10
> the mean of the five numbers is 12
> the range of the five numbers is 8.

Work out the five numbers on the cards.

The sum of all the values is 12 × 5 = 60.
The middle number is 10.
As the mode is 9, the first two numbers must both be 9. As the range is 8, the last number must be 17.
Sum of the four values is 9 + 9 + 10 + 17 = 45
Fifth value is 60 − 45 = 15

9	9	10	15	17

2 Anjali is looking at her homework results from terms 1 and 2. In term 1, she did 10 homeworks, with a mean mark of 60%. In term 2, she did 15 homeworks, with a mean mark of 70%.

Anjali says that because the mean of 60 and 70 is 65 then the mean mark for all of her homeworks is 65%.

Is Anjali correct? You must justify your answer.

Total marks for term 1 is 10 × 60 = 600
Total marks for term 2 is 15 × 70 = 1050
Total number of homeworks is 10 + 15 = 25
So her overall mean mark
$= \dfrac{600 + 1050}{25} = 66\%$
Anjali is not correct as 65% ≠ 66%.

> You must remember the definitions of mean, median and mode.

> The range is the difference between the highest and lowest value, and tells you how spread out the data is.

> To work out the mean here, you require the total of all the data values, which is the total mark for all the tests.

(15) **Exam-style practice** Grade 5

1 Tom is playing a computer game.

At the end of 15 games, his mean score is 72 points per game.

At the end of 14 games, his mean score is 3 points higher.

How many points did Tom score in the 15th computer game?

You must show your working. **[3 marks]**

2 There are 30 students in a class. 18 are boys and 12 are girls. The mean height of the boys is 152 cm. The mean height of the girls is 147 cm.

Work out the mean height of all the students in the class. **[2 marks]**

3 Here is a list of numbers written in order of size.

4   7   x   y

The numbers have a median of 9 and a mean of 12.

What are the values of x and y? **[3 marks]**

Frequency tables

You can use data presented in frequency tables and grouped frequency tables to work out averages.

(5) Finding averages from a frequency table

This table records the numbers of matches in 80 match boxes.

> Draw an extra column in the table to show your working out.

Number of matches, x	Frequency f	x × f
49	23	1127
50	22	1100
51	20	1020
52	15	780
	80	4027

To find the **mean**:

Work out the total number of matches, by multiplying each number by the frequency.

Identify the total of the frequencies. (This is the number of match boxes tested.)

$$\text{Mean} = \frac{\text{total of all values}}{\text{number of values}} = \frac{\sum fx}{\sum f} = \frac{4027}{80} = 50.3$$

The mean number of matches in a box is **50**.

To find the **median**:

The total number of boxes is 80.

The median is the $\frac{80 + 1}{2}$ value, when they are written in order.

This is the 40.5th value. The first 23 values are 49, then the next 22 values are 50 so the 40.5th value must be **50**.

To find the **mode**:

Look for the value that has the highest frequency, which is the one that occurs the most.

The highest frequency is **23**, so the mode is **49**.

(5) Worked example — Grade 6

The table shows information about the number of hours spent on the internet last week.

Number of hours, h	Frequency f	Mid-point x	x × f
$0 \leqslant h < 2$	5	1	5
$2 \leqslant h < 4$	6	3	18
$4 \leqslant h < 6$	4	5	20
$6 \leqslant h < 8$	10	7	70
$8 \leqslant h < 10$	12	9	108
	37		221

> Add columns to the table to show your working. For grouped data always work out the mid-point.

> For grouped data, you can't find the mode, but you can identify the modal class, which is the class with the highest frequency.

> 5 + 6 + 4 = 15 values are less than 6. There are 10 values in the next class interval, so this contains the median.

(a) Write down the modal class.

$8 \leqslant h < 10$

(b) Write down the class interval that contains the median.

The median is the $\frac{37 + 1}{2} = $ 19th value, which is in the class $6 \leqslant h < 8$.

(c) Work out an estimate for the mean number of hours.

$$\text{Mean} = \frac{\sum fx}{\sum f} = \frac{221}{37} = 5.97$$

(d) Explain why your answer to part **(c)** is an estimate.

The data is grouped so we don't know the exact values.

(15) Exam-style practice — Grades 5–6

The table gives information about the temperature, $T\,°C$, at noon in a town for 50 days.

Temperature (T °C)	Frequency
$4 \leqslant T < 8$	6
$8 \leqslant T < 12$	8
$12 \leqslant T < 16$	13
$16 \leqslant T < 20$	21
$20 \leqslant T < 24$	2

(a) Write down the modal class. **[1 mark]**

(b) Write down the class interval which contains the median. **[1 mark]**

(c) Work out an estimate for the mean temperature. **[3 marks]**

(d) What assumption have you made in calculating the mean temperature? **[1 mark]**

Made a start ☐ Feeling confident ☐ Exam ready ☐

Interquartile range

A set of data, arranged in numerical order, may be split into quarters. The **lower quartile** is the value a quarter of the way through the data and the **upper quartile** is the value three-quarters of the way through the data. You need to be able to calculate quartiles and interquartile ranges for data given in a list of stem-and-leaf diagrams.

5 Measures of spread

This diagram shows the different measures of spread.

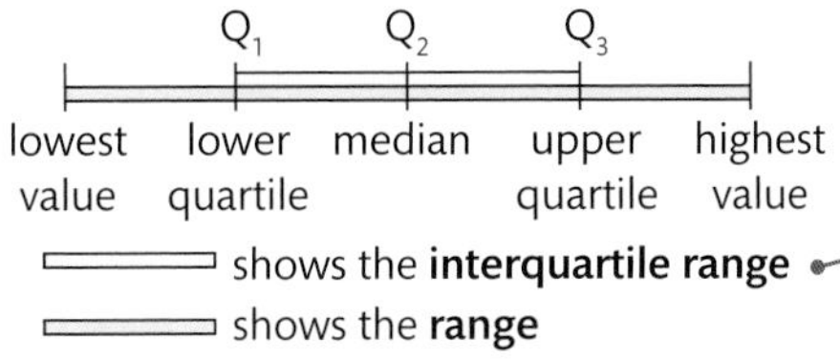

Q_1
Q_2
Q_3
lowest value
lower quartile
median
upper quartile
highest value
shows the **interquartile range**
shows the **range**

The interquartile range represents the middle 50% of the data.

To work out the lower quartile, Q_1

$Q_1 = \frac{1}{4}(n + 1)$th value

To work out the upper quartile, Q_3

$Q_3 = \frac{3}{4}(n + 1)$th value

where n is the number of values and they are listed in numerical order.

Interquartile range = upper quartile – lower quartile

5 Worked example — Grade 5

Work out the interquartile range for these 15 numbers.

17 16 18 14 15 19 17 13 16 15 19 14 13 16 18

13 13 14 (14) 15 15 16 16 16 17 17 (18) 18 19 19

$Q_1 = \frac{1}{4}(15 + 1)$th value = 4th value

$Q_1 = 14$

$Q_3 = \frac{3}{4}(15 + 1)$th value = 12th value

$Q_3 = 18$

Interquartile range = 18 – 14 = 4

You must put the numbers in numerical order before you can work out the quartiles.

5 Worked example — Grade 6

The stem-and-leaf diagram gives information about the heights of 19 plants.

1	6 7
2	1 4 5 6
3	2 3 7 8 8
4	1 2 4
5	4 5 6
6	0 6

Key: 1 | 6 = 16 cm

The first column is the **stem**.

Using this key, you can see that the data, in order, are 16 cm, 17 cm, 21 cm, 24 cm, etc.

Work out the interquartile range.

$Q_1 = \frac{1}{4}(19 + 1)$th value = 5th value

$Q_1 = 25$

$Q_3 = \frac{3}{4}(19 + 1)$th value = 15th value

$Q_3 = 54$

Interquartile range = 54 – 25 = 29 cm

2 Checklist

- To work out the upper quartile, Q_3, find $\frac{3}{4}$ of the $(n + 1)$th value.
- To work out the lower quartile, Q_1, find $\frac{1}{4}$ of the $(n + 1)$th value.
- Remember the formula for the interquartile range: $IQR = Q_3 - Q_1$
- Make sure you know how to read a stem-and-leaf diagram and you understand the key.

10 Exam-style practice — Grades 5–6

Ray counted the letters in each of 31 sentences in a magazine. These numbers show his results.

3 4 5 8 9 10 11 12 13 13 17 18 21 24 26 26 28 28 29 31 32 33 34 37 37 38 39 40 41 42 42

(a) Work out the range. **[1 mark]**

(b) Work out the interquartile range. **[2 marks]**

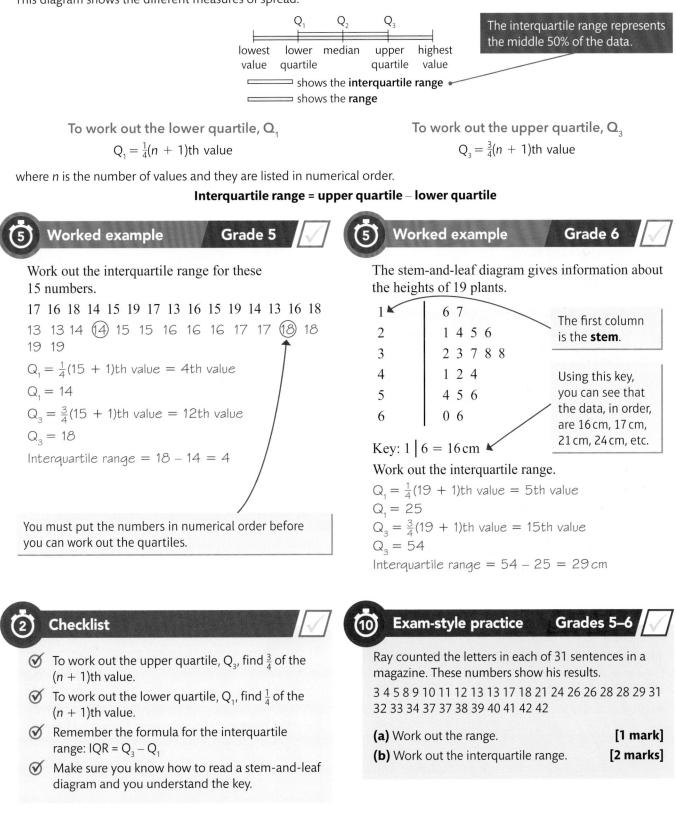

Line graphs

You can use different types of line graph to represent different types of information.

(5) Vertical line graph

This is similar to a bar graph, showing discrete data, except that lines are drawn instead of bars. The height of each line represents a particular frequency.

Scores obtained by quiz teams

(vertical line graph: frequency on y-axis 0–8, score on x-axis 1–10)

From this line graph you can work out the different averages and the spread.

(5) Time series graph

A time series graph displays continuous data. You plot values of a given variable against time. Time is always on the horizontal axis.

Company sales 1986–2000

(time series graph: sales/£million on y-axis 0–70, year on x-axis 86–2000)

From this time series graph you can work out the trend of the data.

(5) Worked example   Grade 5

The vertical line graph shows the numbers of children absent each day over 20 school days.

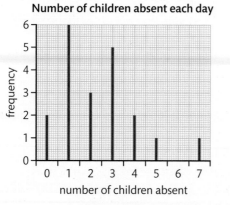

Number of children absent each day

(vertical line graph: frequency on y-axis 0–6, number of children absent on x-axis 0–7)

Work out the mean number of children absent per day.

Mean =

$$\frac{[(0 \times 2) + (1 \times 6) + (2 \times 3) + (3 \times 5) + (4 \times 2) + (5 \times 1) + (6 \times 0) + (7 \times 1)]}{20}$$

$$\text{Mean} = \frac{47}{20} = 2.35$$

When describing trend always use the words 'downwards' or 'upwards'. You just need to look at the shape of the graph and the slope of the line.

(15) Exam-style practice   Grade 5

The table gives information about the profits made by a company, in the summer (S) and in the winter (W), over three years.

	Year 1		Year 2		Year 3	
	S	W	S	W	S	W
Profit (£)	12 000	24 000	18 000	33 000	23 000	39 000

(a) Draw a time series graph for this information.
[3 marks]

(b) Describe the trend in the profits made over these three years. **[1 mark]**

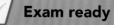

Made a start   Feeling confident   Exam ready

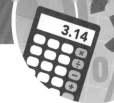

Scatter graphs

A scatter graph is a visual representation of two variables. It can show a trend or a **correlation**. **Strong correlation** is shown by the points being clustered along a line. **Weak correlation** is shown by the points being more openly dispersed along a line.

5 Types of correlation

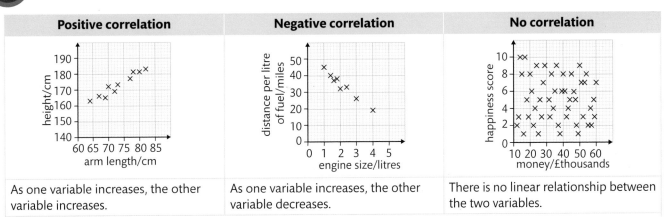

Positive correlation	Negative correlation	No correlation
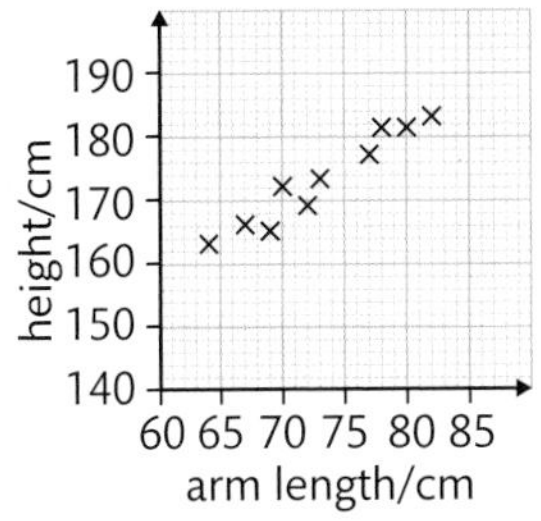	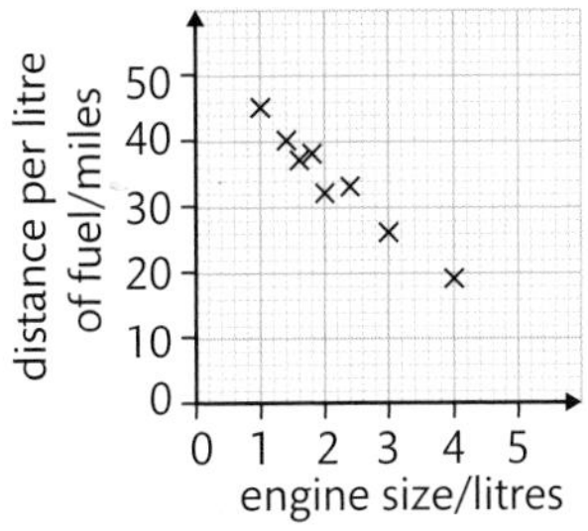	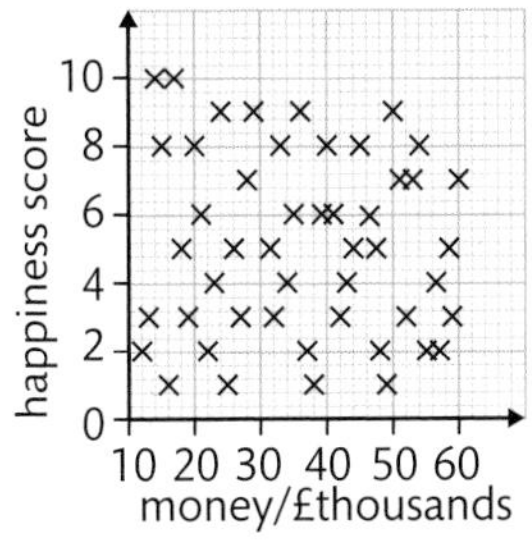
As one variable increases, the other variable increases.	As one variable increases, the other variable decreases.	There is no linear relationship between the two variables.
A **line of best fit** can only be drawn if the scatter graph shows negative or positive correlation.		
Outliers are values that lie outside the 'trend' shown by the rest of the data.		

Making predictions

- **Interpolation** is when you make a prediction when the value is within the given data range.
- The predicted value is reliable.

- **Extrapolation** is when you make a prediction when the value is outside the given data range.
- The predicted value is not reliable.

5 Worked example — Grade 6

A garage sells motorcycles. The scatter graph shows information about the price and age of the motorcycles.

Age of motorcycle against selling price

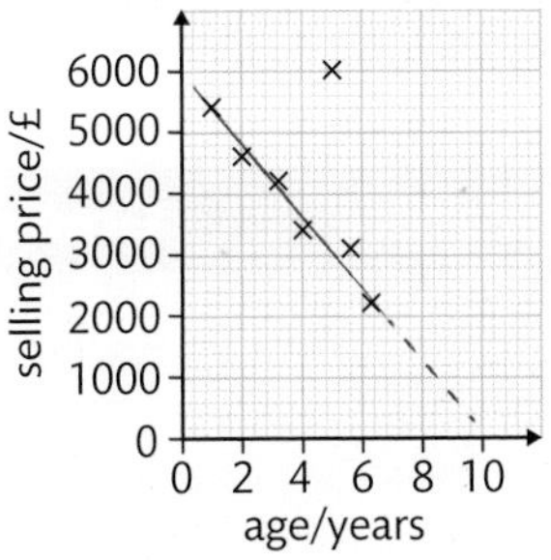

One of the points is an outlier.

(a) Write down the coordinates of the outlier.

(5, 6000)

(b) (i) Draw the line of best fit.

(ii) Describe the correlation.

There is a negative correlation between the selling price of a motorcycle and its age.

A motorcycle is 9 years old.

(c) Estimate the price of this motorbike.

£700

(d) Is this estimate reliable? Give a reason for your answer.

No, this is not a reliable estimate; it is extrapolation, outside the given data range.

The line of best fit should go approximately through the middle of all the points, ignoring any outlier.

15 Exam-style practice — Grade 6

10 students each took a French test and a German test. The table shows their marks.

French marks	44	30	40	50	14	20	32	34	20	45
German marks	48	35	45	54	18	22	36	38	25	50

(a) Show this information on a scatter graph. **[2 marks]**

(b) Describe the relationship, if any, between French marks and German marks. **[1 mark]**

(c) Use a line of best fit to estimate:

(i) the German mark for a student with a French mark of 27

(ii) the French mark for a student with a German mark of 42. **[2 marks]**

Cumulative frequency

A cumulative frequency diagram shows the distribution of a set of continuous data.

⏱5 The shape of a cumulative frequency graph ✓

The shape of the curve gives information about the data.

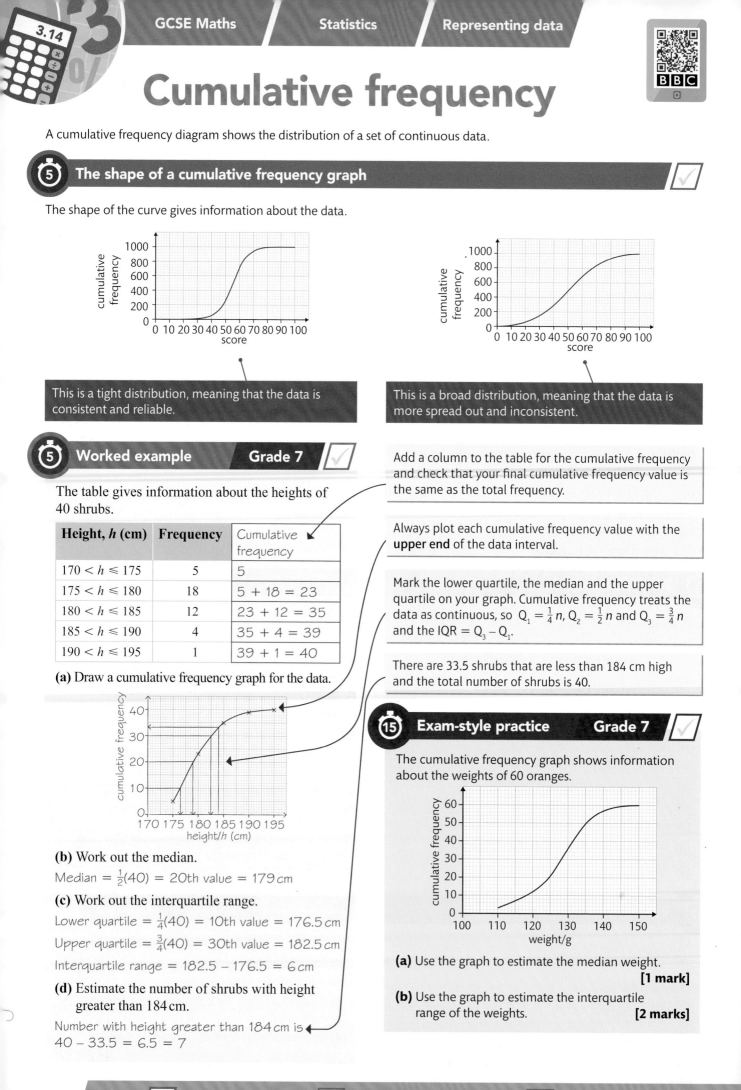

This is a tight distribution, meaning that the data is consistent and reliable.

This is a broad distribution, meaning that the data is more spread out and inconsistent.

⏱5 Worked example — Grade 7 ✓

The table gives information about the heights of 40 shrubs.

Height, h (cm)	Frequency	Cumulative frequency
$170 < h \leqslant 175$	5	5
$175 < h \leqslant 180$	18	$5 + 18 = 23$
$180 < h \leqslant 185$	12	$23 + 12 = 35$
$185 < h \leqslant 190$	4	$35 + 4 = 39$
$190 < h \leqslant 195$	1	$39 + 1 = 40$

(a) Draw a cumulative frequency graph for the data.

(b) Work out the median.

Median = $\frac{1}{2}(40)$ = 20th value = 179 cm

(c) Work out the interquartile range.

Lower quartile = $\frac{1}{4}(40)$ = 10th value = 176.5 cm

Upper quartile = $\frac{3}{4}(40)$ = 30th value = 182.5 cm

Interquartile range = 182.5 − 176.5 = 6 cm

(d) Estimate the number of shrubs with height greater than 184 cm.

Number with height greater than 184 cm is
40 − 33.5 = 6.5 = 7

Add a column to the table for the cumulative frequency and check that your final cumulative frequency value is the same as the total frequency.

Always plot each cumulative frequency value with the **upper end** of the data interval.

Mark the lower quartile, the median and the upper quartile on your graph. Cumulative frequency treats the data as continuous, so $Q_1 = \frac{1}{4}n$, $Q_2 = \frac{1}{2}n$ and $Q_3 = \frac{3}{4}n$ and the IQR = $Q_3 - Q_1$.

There are 33.5 shrubs that are less than 184 cm high and the total number of shrubs is 40.

⏱15 Exam-style practice — Grade 7 ✓

The cumulative frequency graph shows information about the weights of 60 oranges.

(a) Use the graph to estimate the median weight.
[1 mark]

(b) Use the graph to estimate the interquartile range of the weights.
[2 marks]

Made a start ✓   Feeling confident ✓   Exam ready ✓

Box plots

A **box plot** is used to display information about the range, the median and the quartiles in a set of data.

(5) The makeup of a box plot

This diagram shows a box plot. You should always include a scale when you draw a box plot.

Box plots are divided into four parts, separated by five vertical lines. Each part represents 25%.

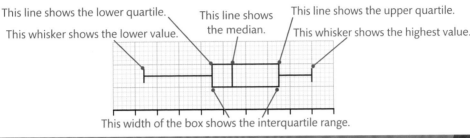

This line shows the lower quartile.   This line shows the median.   This line shows the upper quartile.

This whisker shows the lower value.   This whisker shows the highest value.

This width of the box shows the interquartile range.

(10) Worked example

Grade 7

1 This box plot gives information about the weights of some dogs at a sanctuary.

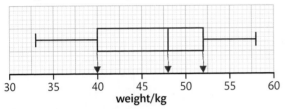

weight/kg

(a) Write down the median.

Median = 48 kg

(b) Work out the interquartile range.

IQR = $Q_3 - Q_1$ = 52 − 40 = 12 kg ◄

There are 60 dogs at the sanctuary.

(c) Write down the number of dogs that weighed 40 kg or less.

40 kg or less = 25% of 60 = 15 dogs

Read the quartiles from the graph and apply the formula. Make sure you read the scale carefully.

2 Rebekah works in a coffee shop. The table gives information about the waiting times, in seconds, for customers served one morning.

Least time	Lower quartile	Median time	Upper quartile	Greatest time
16	28	54	70	94

(a) Draw a box plot for this information.

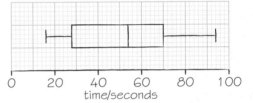

time/seconds

(b) Rebekah said that more than half of her customers waited less than a minute. State, with a reason, whether Rebekah is correct.

She is correct because 60 seconds is above the median.

(15) Exam-style practice

Grade 7

Arif recorded the number of limes on each of 100 lime trees. The incomplete table and box plot give information about his results.

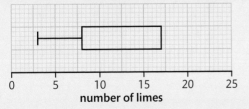

number of limes

	Number of limes
Smallest number	
Lower quartile	8
Median	13
Upper quartile	
Greatest number	24

(a) (i) Use information from the table to complete the box plot.

(ii) Use information from the box plot to complete the table.

[3 marks]

Some of these 100 lime trees have 8 or more limes on them.

(b) Write down the number of lime trees with 8 or more limes on them.

[2 marks]

Histograms

A **histogram** is used to display grouped continuous data. You can use a histogram to display data when the class widths are not all the same.

⑤ About histograms ✓

A histogram can be used to represent information with equal or unequal class widths. The class width is the difference between the upper and lower class boundaries. The two main characteristics of a histogram are

- there are no gaps between the bars
- the area of each bar is proportional to the frequency of that data value.

The vertical axis on a histogram is always labelled as **frequency density**. To work out frequency density you use the formula:

$$\text{frequency density} = \frac{\text{frequency}}{\text{class width}}$$

> You are drawing a histogram so you need to work out the frequency density for each class. If you have to choose a scale for your vertical axis, it's a good idea to work out all of the frequency densities first so you know what the largest value is.

> For part **(b)** first work out the frequency density, as shown in the third column in the table.

Problem solving ⚙

Draw a line on your graph to help answer part **(b)**.

To estimate the number of plants you need to add the frequency values of the plants that fall to the left of this line. To find the frequency for $12 < w \le 15$, multiply the frequency density of $12 < w \le 16$ by the new class width, 3.

⑤ Worked example — Grade 8 ✓

The table shows some information about the weight, w kg, of tomatoes produced by 62 tomato plants.

Weight, w kg	Frequency	Frequency density
$5 < w \le 10$	8	$8 \div 5 = 1.6$
$10 < w \le 12$	10	$10 \div 2 = 5$
$12 < w \le 16$	14	$14 \div 4 = 3.5$
$16 < w \le 20$	18	$18 \div 4 = 4.5$
$20 < w \le 25$	12	$12 \div 5 = 2.4$

(a) Draw a histogram to show this information.

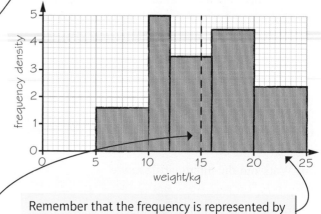

> Remember that the frequency is represented by the area of a rectangular bar.

(b) Work out an estimate for the number of tomato plants that produced 15 kg or less of tomatoes.

Number of tomato plants $= 8 + 10 + (3 \times 3.5)$
$$= 8 + 10 + 10.5$$
Number of tomato plants $= 28.5$

⑮ Exam-style practice — Grade 8 ✓

The table shows some information about the time, in minutes, taken by some adults to travel to work in one week.

(a) Use the table to draw a histogram. **[2 marks]**

(b) Work out an estimate for the median time. **[2 marks]**

Time (t minutes)	Frequency
$0 < t \le 20$	20
$20 < t \le 30$	30
$30 < t \le 40$	45
$40 < t \le 60$	60
$60 < t \le 100$	48

Made a start ✓ Feeling confident ✓ Exam ready ✓

Frequency polygons

A **frequency polygon** is a line graph that describes the shape of a distribution.

5 Frequency polygons

To construct a frequency polygon, draw a histogram and then join the midpoints of the tops of the bars of the histogram to form a line graph.

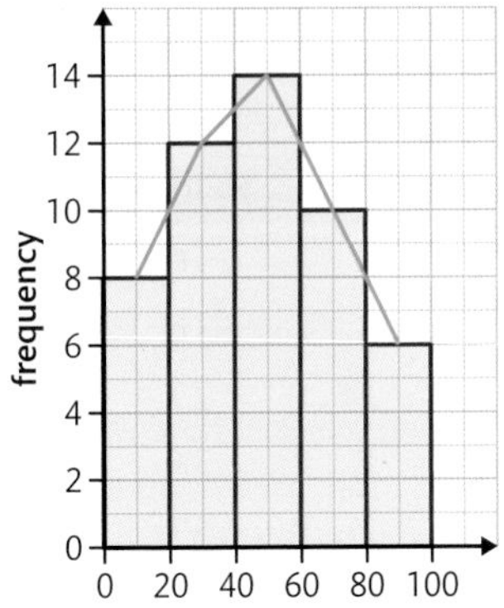

Alternatively, just plot the midpoints of the class intervals and join them up to make a line graph.

5 Worked example — Grade 6

The table shows some information about the weights (w grams) of 70 pears.

Weight (w grams)	Frequency
$100 \leqslant w < 110$	7
$110 \leqslant w < 120$	11
$120 \leqslant w < 130$	18
$130 \leqslant w < 140$	24
$140 \leqslant w < 150$	10

(a) Draw a frequency polygon to show this information.

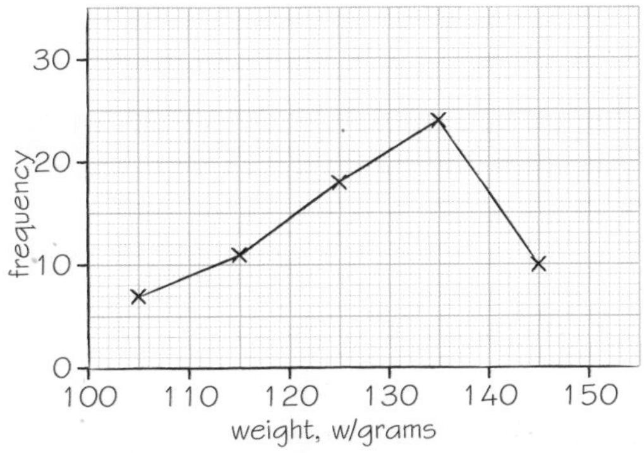

You could have drawn a histogram and then joined the midpoints of the tops of the bars with lines.

(b) Write down the modal class interval.

$130 \leqslant w < 140$

The modal class interval is the one with the highest frequency.

5 Worked example — Grade 6

The table shows some information about the ages, in years, of 80 people.

Age (a years)	Frequency
$20 \leqslant a < 30$	15
$30 \leqslant a < 40$	28
$40 \leqslant a < 50$	18
$50 \leqslant a < 60$	11
$60 \leqslant a < 70$	8

(a) Write down the class interval that contains the median.

Median $= \frac{1}{2}n = \frac{1}{2}(80) = 40$th value

The 40th value lies in $30 \leqslant a < 40$

This **incorrect** frequency polygon has been drawn from the information in the table.

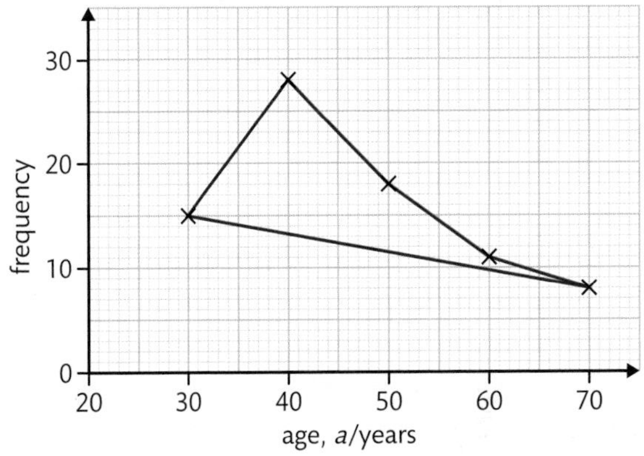

(b) Write down **two** things that are wrong with this incorrect frequency polygon.

The points should be plotted at the mid-interval values.

The last point should not be joined to the first, to close the polygon.

15 Exam-style practice — Grade 6

60 students take a Mathematics test that is marked out of 50. This table shows information about the students' marks.

Maths mark	0–10	11–20	21–30	31–40	41–50
Frequency	3	14	16	18	9

On a grid, draw a frequency polygon to show this information. **[2 marks]**

Analysing data

Averages and measures of spread can be used to analyse and compare data.

⑤ **Comparing data** ✓

You need to be able to compare two sets of data by using the appropriate average and measure of spread.

You need to compare one statistic from each column in this table.

Averages	Measures of spread
Mean	Range
Median	Interquartile range

⑩ **Worked example** **Grade 6** ✓

1 These box plots show information about the times, in minutes, some boys and some girls took to do the same jigsaw puzzle.

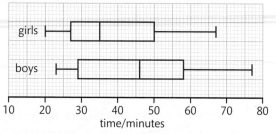

Compare the distributions of the girls' times and the boys' times.

The median time for the boys is 46, which is greater than the median time for the girls, which is 35. So, on average girls completed the puzzle more quickly.

The interquartile range and the range show that the boys' times are more spread out than the girls' times.

2 Harry and Suki recorded the times, in minutes, they exercised each morning for five days.

Harry	Suki
12 16 12 13 19	16 11 16 18 20

Who exercised for the longer time? You must justify your answer.

$$\text{Mean for Harry} = \frac{12 + 16 + 12 + 13 + 19}{5}$$
$$= 14.4 \text{ minutes}$$

$$\text{Mean for Suki} = \frac{16 + 11 + 16 + 18 + 20}{5}$$
$$= 16.2 \text{ minutes}$$

Suki exercised for longer as her mean is greater.

⑤ **Worked example** **Grade 6** ✓

The owners of a village hall recorded the ages of some people at parties on a Friday night and on a Saturday night. The table shows information about the ages of people on the Friday night and on the Saturday night.

Age	Friday	Saturday
Lowest	16	4
Lower quartile	18	8
Median	27	13
Upper quartile	36	30
Highest	57	62

(a) What type of diagram could you draw to represent the information for each day?

Box plot

(b) Compare the distribution of the ages of the people at the party on Saturday with the distribution of the ages of the people at the party on Friday.

The median age on Friday was greater than the median age on Saturday so on average the people on Friday are older than the people on Saturday.

The range of ages on Friday is smaller than the range of ages on Saturday.

The median is interpreted in context.

You must interpret the statements in context.

⑩ **Exam-style practice** **Grade 6** ✓

Joanne is recording the times, in minutes, that some people spent at a motorway cafe during the day and during the night. Her notes show this information.

Compare the distribution of the time spent during the day to the distribution of the time spent during the night. **[4 marks]**

Daytime
6 10 12 15 16
21 23 26 28 28
29 31 33 36 45

Night time
Least = 7
LQ = 13
Median = 21
UQ = 36
Highest = 56

✓ **Made a start** ✓ **Feeling confident** ✓ **Exam ready**

Statistics

Read the exam-style question and worked solution, then practise your exam skills with the question at the bottom of the page.

⑩ Worked example   Grade 6 ✓

North Farm grows plums for markets. A market inspector weighed 60 plums. The cumulative frequency graph shows some information about these weights.

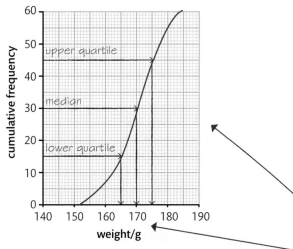

(a) Use the graph to work out an estimate for the median weight. Circle your answer.

168   (170)   166   172

The 60 plums from North Farm had a minimum weight of 152 grams and a maximum weight of 184 grams.

(b) Use this information and the cumulative frequency graph to draw a box plot for the 60 plums from North Farm.

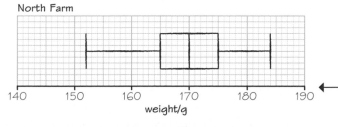
North Farm

The market inspector then visited South Farm. Again, he weighed 60 plums. He drew this box plot for his results.

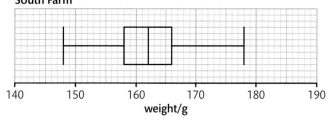
South Farm

(c) Compare the distributions of the weights.

The median of North Farm is greater than the median of South Farm so on average the plums from North Farm weigh more than the plums from South Farm.
The range and IQR of North Farm are greater than the range and IQR of South Farm (more variation).

> To identify the median, draw an arrow at the value of $\frac{1}{2}n$ and then read off the value on the horizontal axis.

> To identify the lower quartile, draw an arrow at the value of $\frac{1}{4}n$ and then read off the value on the horizontal axis. To identify the upper quartile, draw an arrow at the value of $\frac{3}{4}n$ and then read off the value on the horizontal axis.

> To draw a box plot you need five values: lowest, lower quartile, median, upper quartile and highest.

⑩ Exam-style practice   Grade 6 ✓

A teacher recorded the times, in minutes, for 15 boys to complete a puzzle. Here are his results:

4 8 9 17 20 21 25 27 29 30 33 34 38 43 46

(a) On the grid, draw a box plot for this information.

[3 marks]

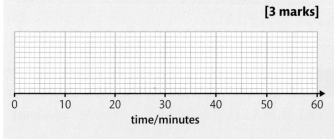

The teacher recorded the time, in minutes, for 15 girls to complete the same puzzle. The box plot shows some information about these times.

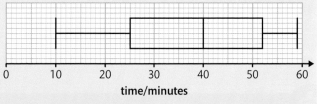

(b) Compare the amount of time the boys spent on the puzzle with the amount of time the girls spent on the puzzle. **[2 marks]**

Problem solving strategies

In your exam, questions might require you to make connections between different parts of mathematics, interpret results and evaluate methods within a certain context, or think about problems as a series of mathematical processes. This is known as **problem solving**.

⑤ How to answer problem-solving questions

Ask yourself these **five** sets of questions.

① What do I have to do?

Have I identified the type of problem?

Have I read through the problem carefully?

Have I thought about the problem in my own words?

Have I reviewed my thoughts and ideas?

② What information do I need?

Have I looked at the key words used in the question?

Have I understood the mathematical and statistical language used?

What is the style of the question – describe, show, solve, prove, or state?

How do I respond to these key mathematical words?

③ What mathematics can I do?

What topic(s) do I need to recall?

What formulae do I need?

What steps are required to reach the answer?

What would a sensible answer look like?

④ Is my solution correct?

Have I carried out my plan logically?

Have I checked my working?

Are my answers sensible?

⑤ Have I completed everything?

Have I answered every part of the question?

Have I given the information that was being asked for?

Have I shown all my working?

Have I given units with my answer?

⑤ Worked example — Grade 6

Anjali has an empty bag. She puts some white beads and some red beads into the bag. The ratio of red to white is $1:3$.

Anjali takes, at random, two beads from the bag. The probability that she takes two red beads is $\frac{1}{19}$.

How many beads in total did she put into the bag?

What is my plan to solve the problem?

I need to draw a tree diagram.

There are x red beads so there are $3x$ white beads.

Total number of beads is $x + 3x$.

I need to label the first red branch as $\frac{x}{4x}$

I need to label the second red branch as $\frac{x-1}{4x-1}$

I need to multiply along the branches $\frac{x}{4x} \times \frac{x-1}{4x-1}$

I need to set up an equation $\frac{x}{4x} \times \frac{x-1}{4x-1} = \frac{1}{19}$

I need to solve the equation to work out the value of x.

⑤ Worked example — Grade 6

These are the first three terms of a Fibonacci-type sequence.

$a \quad b \quad a+b$

The third term is 5 and the seventh term is 34.

What are the values of a and b?

What is my plan to solve the problem?

I need to work out the seventh term. The fourth to seventh terms are:

$a+2b \quad 2a+3b \quad 3a+5b \quad 5a+8b$

I need to equate the third algebraic term to 5 so

$a + b = 5$

I need to equate the seventh algebraic term to 34 so

$5a + 8b = 34$

I need to solve two simultaneous equations to work out a and b.

$5a + 8b = 34$
$a + b = 5$

⑩ Exam-style practice — Grades 6–9

Using the five-point plan for problem solving, solve the two problems set on this page. **[4 marks]**

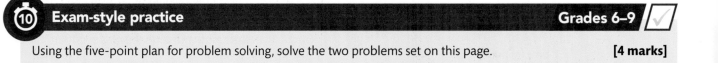

✓ Made a start   ✓ Feeling confident   ✓ Exam ready

Solving number problems

Number skills are often tested in particular contexts, such as problems with money.

⏱ Worked example — Grade 6

Dale bought some shares for <u>£5400</u>. The value of his shares <u>increased by 2.5% in the first year</u>. Their value <u>decreased by 1.5% in the second year</u>. Then their value <u>increased by x% in the third year</u>. At the <u>end of the third year the value of the shares was £5610.08</u>.

Laura has invested <u>£6000</u> in a savings account for <u>three years</u>. The account pays <u>compound interest</u> at annual rates of

    <u>3.5% for the first year</u>

    <u>y% for the second year</u>

    <u>y% for the third year</u>

There is a total amount of <u>£6554.97 in the savings account at the end of three years</u>.

Who earned more interest, <u>proportionally</u>, in the third year? You must show your working.

> You need to compare their returns on their investments.

Dale

$$5400 \times 1.025 \times 0.985 \times \left(1 + \frac{x}{100}\right) = 5610.08$$

$$5451.975 \times \left(1 + \frac{x}{100}\right) = 5610.08$$

$$1 + \frac{x}{100} = \frac{5610.08}{5451.975}$$

$$x = 2.9\%$$

> To work out an original amount, divide by the multiplier.

Laura

$$6000 \times 1.035 \times \left(1 + \frac{y}{100}\right)^2 = 6554.97$$

$$6210 \times \left(1 + \frac{y}{100}\right)^2 = 6554.97$$

$$\left(1 + \frac{y}{100}\right)^2 = \frac{6554.97}{6210}$$

$$1 + \frac{y}{100} = \sqrt{\frac{6554.97}{6210}}$$

$$y = 2.7\%$$

> To remove a square from one side, take the square root of the other side.

Dale earned a higher rate of interest in the third year as 2.9% is greater than 2.7%.

⚙ Problem solving

❶ Do I understand the problem?

The question is asking you to work out the percentage of interest earned by both Dale and Laura, and then to state which of those values is higher.

❷ What information do I need?

You need all the underlined information.

❸ What maths do I know?

- A simple way to work out compound interest is to work out the multiplier, using $\frac{100\% + x\%}{100\%}$ where x% is the given percentage.

- The formula for compound interest is initial amount × (multiplier)^n.

❹ Is my solution correct?

The answer is sensible considering the initial amount of interest given and the working is all correct.

❺ Have I completed everything?

Re-read the question. You need to show the amount of interest for both Dale and Laura as percentages and decide which is the higher value.

> Depreciation takes place when the value of an item decreases.
>
> To calculate the multiplier for depreciation, work out $\frac{100 - x}{100}$ (where x% is the given percentage).

⏱ Exam-style practice — Grade 6

Ani invests £12 000 in an account for three years.

The account pays 3.5% compound interest per annum.

Ani has to pay 40% tax on the interest earned each year. This tax is taken from the account at the end of each year.

How much money will Ani have in the account at the end of the three years? **[4 marks]**

Solving proof problems

If you are asked to prove an identity, you can use algebraic techniques.

[2] Identities and proof

An identity is an equation which is true for all possible numerical values of its variables.

The sign used for an identity is $\equiv$.

An example of an identity is

$(n-2)^2 + (n-6) \equiv n^2 - 3n - 2$

The left-hand side must be equal to the right-hand side.

An algebraic proof shows that an identity is true for all values of n.

[2] Constructing proof

When constructing an algebraic proof, you can use the following algebraic expressions to represent different types of numbers.

Number	Algebraic expression
number	n
even	$2n$
odd	$2n + 1$
square	n^2
cube	n^3

[2] Worked example — Grade 5

Prove that $(n-3)^2 - (n-5) \equiv (n-3)(n-4) + 2$.

$(n-3)^2 - (n-5) \equiv (n-3)(n-3) - n + 5$

$\equiv n^2 - 3n - 3n + 9 - n + 5$

$\equiv n^2 - 7n + 12 + 2$

$\equiv (n-3)(n-4) + 2$

Multiply out the brackets in the expression on the left and then simplify the expressions.

[2] Worked example — Grade 7

Prove algebraically that the difference between the squares of any two consecutive even numbers is always a multiple of 4.

$(2n+2)^2 - (2n)^2 = 4n^2 + 4n + 4n + 4 - 4n^2$

$= 8n + 4$

$= 4(2n + 1)$

$2n$ is an even number. The next even number will be $2n + 2$. The squares of these numbers are $(2n)^2$ and $(2n+2)^2$.

[5] Worked example — Grade 5

Prove that $\frac{1}{4}(2n+1)(n+4) - \frac{1}{4}n(2n+1) \equiv 2n + 1$.

$\frac{1}{4}(2n+1)(n+4) - \frac{1}{4}n(2n+1)$

$\equiv \frac{1}{4}(2n^2 + 8n + n + 4) - \frac{1}{4}(2n^2 + n)$

$\equiv \frac{1}{4}(2n^2 + 8n + n + 4 - 2n^2 - n)$

$\equiv \frac{1}{4}(8n + 4)$

$\equiv 2n + 1$

Multiply out the first pair of brackets before multiplying the result by $\frac{1}{4}$.

[5] Worked example — Grade 7

Prove algebraically that when you divide the sum of the squares of any three consecutive odd numbers by 12, there is always a remainder of 11.

$(2n+1)^2 + (2n+3)^2 + (2n+5)^2$

$= 4n^2 + 4n + 1 + 4n^2 + 12n + 9 + 4n^2 + 20n + 25$

$= 12n^2 + 36n + 35$

$= 12n^2 + 36n + 24 + 11$

$= 12(n^2 + 3n + 2) + 11$

Consecutive odd numbers are $2n + 1$, $2n + 3$ and $2n + 5$.

[15] Exam-style practice — Grade 5–7

1. Prove that $\frac{1}{2}(n+1)(n+2) - \frac{1}{2}n(n+1) \equiv n + 1$. **[3 marks]**

2. Prove that $(5n+1)^2 - (5n-1)^2$ is a multiple of 4 for all positive values of n. **[3 marks]**

3. Prove that $(n+1)^2 - (n-1)^2 + 1$ is always odd for positive values of n. **[3 marks]**

4. Prove that, for any three consecutive numbers, the difference between the squares of the first and last numbers is four times the middle number. **[4 marks]**

Solving geometric problems

Geometric skills are often tested in particular contexts, such as problems with angles and using similar triangles to find lengths of sides.

(10) Worked example — Grade 6

The diagram shows a metal frame ABCDEFGHJ, which is to be used for the construction of a ramp.

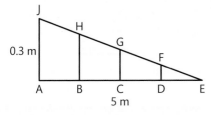

The line AE is horizontal and the line AJ is vertical.

JA, BH, CG and DF are parallel.

AB = BC = CD = DE

AE = 5 m and AJ = 0.3 m

The legal requirement for the angle AEJ of the ramp is a maximum of 3.8°.

(a) Does this ramp meet the legal requirement?

opp = 0.3 and adj = 5

SOH CAH ⟨TOA⟩

$$\tan x = \frac{opp}{adj}$$

$$\tan x = \frac{0.3}{5}$$

$$x = \tan^{-1}\left(\frac{0.3}{5}\right)$$

$$x = 3.43°$$

The angle meets the legal requirement.

> Write down SOH CAH TOA and then, from what you have written down, choose the correct ratio.

> To work out the angle you need to use $\tan^{-1}$.

(b) Show that BH is 22.5 cm.

From similar triangles:

$$\frac{y}{0.3} = \frac{3.75}{5}$$

$$y = \frac{3.75}{5} \times 0.3$$

$$y = 0.225 \, m$$

$$y = 22.5 \, cm$$

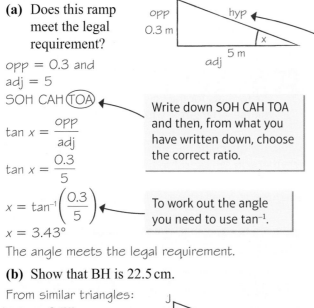

(5) Problem solving

1 Do I understand the problem?

The question is asking you to apply trigonometry to right-angled triangles.

2 What information do I need?

You need to know that you are dealing with right-angled triangles.

3 What maths do I know?

- The trigonometric relationships for right-angled triangles are SOH CAH TOA.
- The framework is a series of similar triangles.

4 Is my solution correct?

The answers are sensible considering the diagram given, and the working is all correct.

5 Have I completed everything?

Re-read the question. You need to show whether the ramp meets the legal requirement and that the length in part **(b)** is correct.

> Always label the sides and then write down what you know and what you are looking for.

(10) Exam-style practice — Grade 6

The diagram shows part of the roof of a bungalow. The roof is in the shape of a square-based pyramid *ABCDE*.

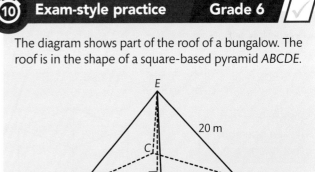

AE = BE = CE = DE = 20 m

AB = 25 m

As part of the design of the roof, angle *DEB* must not exceed 125°. Is the design correct? You must show your working. **[4 marks]**

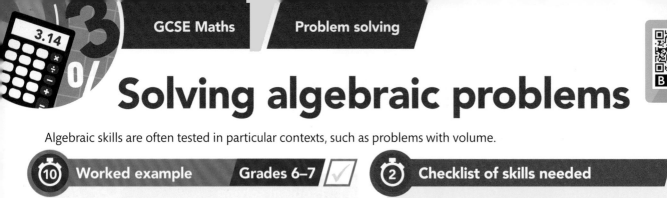

Solving algebraic problems

Algebraic skills are often tested in particular contexts, such as problems with volume.

🕙 Worked example   Grades 6–7 ✓

An open box is made by cutting a square of side 2 cm from each corner of a square metal sheet and then folding the sides up. The length of the metal sheet is x cm. The volume of the box is 200 cm³.

(a) Show that $x^2 - 8x - 84 = 0$.   **[4 marks]**

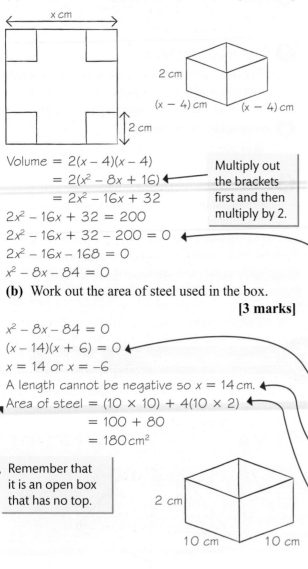

Volume $= 2(x - 4)(x - 4)$
$= 2(x^2 - 8x + 16)$ ← Multiply out the brackets first and then multiply by 2.
$= 2x^2 - 16x + 32$
$2x^2 - 16x + 32 = 200$
$2x^2 - 16x + 32 - 200 = 0$ ←
$2x^2 - 16x - 168 = 0$
$x^2 - 8x - 84 = 0$

(b) Work out the area of steel used in the box.   **[3 marks]**

$x^2 - 8x - 84 = 0$
$(x - 14)(x + 6) = 0$ ←
$x = 14$ or $x = -6$
A length cannot be negative so $x = 14$ cm. ←
Area of steel $= (10 \times 10) + 4(10 \times 2)$ ←
$= 100 + 80$
$= 180$ cm²

Remember that it is an open box that has no top.

2️⃣ Checklist of skills needed ✓

- ☑ Writing measurements as algebraic expressions
- ☑ Working out the volume of a cuboid
- ☑ Multiplying out brackets
- ☑ Setting up equations
- ☑ Writing expressions as quadratics
- ☑ Solving quadratic equations
- ☑ Working out the surface area of a cuboid

2️⃣ Problem solving ⚙️ ✓

1 **Do I understand the problem?**
You need to use algebraic skills to work out the dimensions and the surface area of a box.

2 **What information do I need?**
- You need to know that four squares, all the same size, are cut from the corners of a metal sheet. It is then folded to make a box.
- You need to sketch a diagram of the net and the open box and write down the measurements that you know.

3 **What maths do I know?**
The formula for the volume of a cuboid is volume = length × width × height.

When the quadratic equation is set up, always check whether you can cancel so that it is in its simplest form.

You need to recall how to solve quadratic equations.

You need to know that when questions are about measurements the negative value is ignored.

You need to recall that the surface area of a cuboid is the sum of the areas of all the sides.

2️⃣ Problem solving ⚙️ ✓

4 **Is my solution correct?**
The answers are sensible considering the volume of the box given, and the working is all correct.

5 **Have I completed everything?**
Re-read the question. It asks for a quadratic equation so your answer should be in this form.

🕙 Exam-style practice   Grades 6–7 ✓

The diagram shows a rectangle ABCD.
The perimeter of the rectangle is 68 cm.
The diagonal AC is 26 cm.

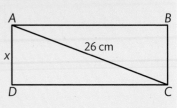

(a) Show that
$x^2 - 34x + 240 = 0$.   **[4 marks]**

(b) Work out the length AB.   **[3 marks]**

Solving statistical problems

Statistical skills are often tested in particular contexts, such as problems with Venn diagrams and conditional probabilities.

 Worked example | Grade 6

80 people were asked which of three sports they watched on television. Here are the results.

38 people watched tennis.

29 people watched athletics.

35 people watched swimming.

17 people watched both tennis and athletics.

18 people watched both tennis and swimming.

16 people watched both athletics and swimming.

11 people watched all three sports.

(a) Draw a Venn diagram to show this information.

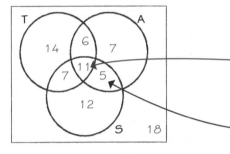

One of the 80 people is selected at random.

(b) Work out the probability that the person does not watch any of the three sports.

$$P(\text{no sport}) = \frac{18}{80} = \frac{9}{40}$$

(c) Given that a person watches tennis, what is the probability that this person also watches swimming?

$$P(\text{swimming} \mid \text{tennis}) = \frac{18}{38} = \frac{9}{19}$$

(d) Given that a person watches at least one of the sports, what is the probability that this person watches all three?

$$P(\text{all 3} \mid \text{at least 1 sport}) = \frac{11}{62}$$

 Problem solving

1 Do I understand the problem?

The question is asking you to draw and label the Venn diagram and use the completed Venn diagram to work out some probabilities.

2 What information do I need?

You need to know how to use the information given.

3 What maths do I know?

You need to know how to:

- draw and fill in a Venn diagram
- work out probability from a Venn diagram
- use Venn diagrams to work out conditional probability.

Fill in the numbers on the Venn diagram by starting with the overlap and then working outwards.

16 watched athletics and swimming but you already have 11 who watched all three, so you need another 5 who have watched just athletics and swimming. In a similar way, you obtain 7 and 6 in the other intersections.

Given that they watched tennis, look only at the tennis circle and you can see that 18 also watched swimming.

 Problem solving

4 Is my solution correct?

The answers are a sensible proportion of the total number of people.

5 Have I completed everything?

Re-read the question. Parts **(b)**, **(c)** and **(d)** all ask for a probability, so write your answer as a fraction and with the correct notation.

 Exam-style practice | Grade 6

A sixth form has 100 students in Year 12.

21 students study Biology.

27 students study Chemistry.

32 students study Physics.

9 students study both Biology and Chemistry.

12 students study both Chemistry and Physics.

7 students study both Physics and Biology.

3 students study all three of these subjects.

(a) Draw a Venn diagram to represent this information. **[4 marks]**

A Year 12 student at the college is selected at random.

(b) Work out the probability that the student studies only Chemistry. **[1 mark]**

(c) Given that the student studies Biology, find the probability that this student also studies Physics. **[2 marks]**

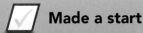

Answers

Page 1 Fractions, decimals and percentages

1 £900

2 3

3 $x = \frac{446}{555}$

Page 2 Manipulating fractions

1 54

2 6

Page 3 Percentage change

1 12.5%

2 £1236.42

Page 4 Reverse percentages

1 2.5 kg

2 In 2015, Zak earned £32 163.46 and Zoe earned £31 940.37 so Zak's commission was greater in 2015.

Page 5 Growth and decay

1 $n = 4$

2 0.82 m

Page 6 Estimation

1 **(a)** 20 **(b)** 3600 **(c)** 150

2 **(a)** 1000

(b) 160

Page 7 Upper and lower bounds

1 16.939

2 He did not exceed the speed limit, because the upper bound of his speed is 79.9 km/h.

Page 8 Accuracy and error

1 $89.5 \leqslant t < 90.5$

2 48

Page 9 Factors and primes

1 **(a)** 12 **(b)** 216

2 **(a)** $2^3 \times 7$ **(b)** 14 **(c)** 392

Page 10 Standard form

(a) 4.5×10^7 km **(b)** 3.4×10^4 km

Page 11 Surds

1 **(a)** $3\sqrt{2}$ **(b)** $5\sqrt{2}$

(c) $3\sqrt{5}$ **(d)** $28\sqrt{2}$

2 $3\sqrt{11}$

Page 12 Exam skills: Number

1 $1.53 \leqslant s < 1.59$

2 2.3 seconds because the upper and lower bound are the same when rounded to 2 significant figures.

Page 13 Algebraic expressions

1 $6y^2 + 15x^3 + 6$

2 **(a)** $-11y - 17$ **(b)** $x^3 - x$

3 $5(2x + 1)(2x - 1)$ **4** $2x(2x + 1)$

Page 14 Algebraic formulae

1 10.1

2 **(a)** $12x + 10y = 200, 10y = 200 - 12x, y = 20 - 1.2x$

(b) $A = 4x \times 2y = 4x \times 2(20 - 1.2x) = 4x(40 - 2.4x)$
$= 160x - 9.6x^2$

Page 15 Laws of indices

1 p^9

2 **(a)** x^8 **(b)** $12x^3y^5$

3 **(a)** m^6 **(b)** m^{15} **(c)** $12w^8y^4$

(d) $8x^4y^7$ **(e)** $64x^9y^{12}$

4 $n = 3$

Page 16 Combining indices

1 **(a)** 4 **(b)** $\frac{1}{5}$

2 **(a)** $\frac{1}{8}$ **(b)** $-\frac{3}{2}$

3 **(a)** $\frac{1}{2}$ **(b)** $\frac{64}{27}$

4 $n = 2.5$ **5** $y = 2.5$

Page 17 Simple linear equations

1 **(a)** $x = \frac{21}{6} = 3.5$ **(b)** $x = \frac{3}{2} = 1.5$
(c) $x = 4$

2 Ann is 24, Ben is 48 and Carl is 20.

Page 18 Linear equations and fractions

1 $x = -3$ **2** $x = \frac{7}{10}$

Page 19 Simultaneous equations

1

$x = 3.7$ and $y = 1.8$

2 $x = \frac{2}{3}$ and $y = -\frac{3}{2}$ **3** Adult £7.50 and child £3.00

Page 20 Quadratic equations

1 **(a)** $x = \frac{5}{3}$ and $x = -\frac{5}{3}$ **(b)** $x = 4$ and $x = -\frac{1}{3}$

2 **(a)** $\frac{1}{2}(x + 3 + x + 1)(x - 4) = 55$

$\qquad \frac{1}{2}(2x + 4)(x - 4) = 55$

$\qquad (x + 2)(x - 4) = 55$

$\qquad x^2 - 2x - 63 = 0$

(b) Solutions are $x = -7$ and $x = 9$, so $x = 9$ because x must be positive.

Page 21 Mixed simultaneous equations

(a) (2.5, 12.5) and (−4, 32)   **(b)** (6, 0) and (−3.6, −4.8)

(c) (2.87, −0.87) and (−0.87, 2.87)

Page 22 Completing the square

1 $a = 4$ and $b = -11$ **2** $x = 3 \pm \sqrt{\frac{22}{3}}$

3 $a = 3$ and $b = -4$

Page 23 The quadratic formula

1 $x = -1.55$ and $x = 0.22$

2 **(a)** 2 solutions ($x = 0.56$ and $x = -0.89$)

(b) no solutions

3 $4x^2 - 13x + 5 = 0$

Page 24 Linear inequalities

1 5, 6 and 7   **2** 6, 7 and 8

3 **(a)** $x > 3$ **(b)** 4

Page 25 Quadratic inequalities

1 **(a)** $x < -3$ and $x > 3$

(b)

2 **(a)** $x < 4$ and $x > 5$ **(b)** $x < -3$ and $x > 6$
(c) $-\frac{4}{3} < x < 2$

3 **(a)**   **(b)** $x \leqslant -1$ and $x \geqslant 3$

Page 26 Arithmetic sequences

1 $11 - 3n$

2 **(a)** Because it does not end in 3 or 8.
(b) $5n - 2$

Page 27 Quadratic sequences

1 $2n^2 + 4n - 3$

2 **(a)** Second differences are equal (2).   **(b)** $n^2 + 5$

Page 28 Sequence problems

1 $a = 3$ and $b = 5$

2 **(a)** The first six terms are a, b, $a + b$, $a + 2b$, $2a + 3b$, $3a + 5b$

(b) $a = 1$ and $b = 3$

3 **(a)** No (because n would need to be 13.5).
(b) $u_n = u_{n-1} + 4$

Page 29 Drawing straight-line graphs

1   **2** $y = -2x + 5$

Page 30 Equations of straight lines

1 $y = -3x + 11$ **2** $y = \frac{1}{2}x - 2$

3 $y = \frac{3}{4}x + 2$

Page 31 Parallel and perpendicular lines

1 **(a)** $y = 3x$ **(b)** $y = -\frac{1}{3}x + 5$

2 $y = -\frac{1}{2}x + 2$

3 Gradient of line A $= \frac{3}{2}$, gradient of line B $= 2$. They are not parallel so they must intersect.

Page 32 Quadratic graphs

(a)

x	−1	0	1	2	3	4
y	6	2	0	0	2	6

(b) $y = x^2 - 3x + 2$

(c) $x = 1$ and $x = 2$

Page 33 Cubic and reciprocal graphs

1 **(a)**

x	−2	−1	0	1	2
y	−13	−4	−1	2	11

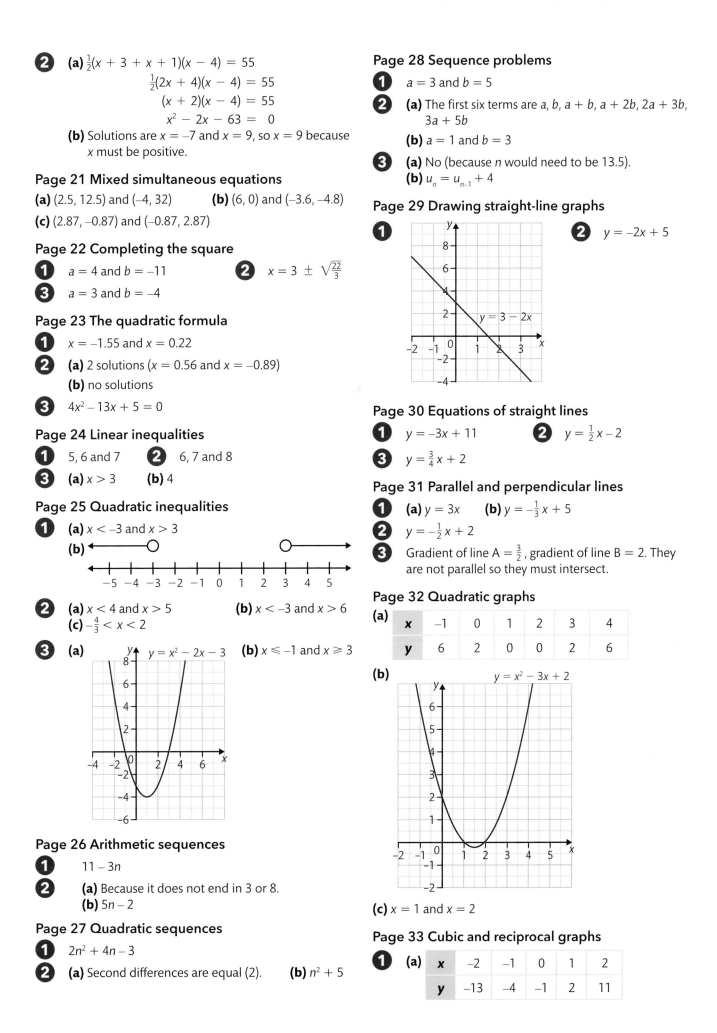

(b)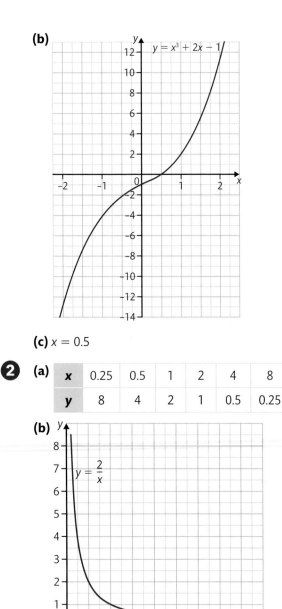

(c) $x = 0.5$

2 **(a)**

x	0.25	0.5	1	2	4	8
y	8	4	2	1	0.5	0.25

(b)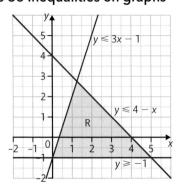

Page 34 Real-life graphs

(a) -0.625 **(b)** A, because it has a steeper gradient.

Page 35 Trigonometric graphs

1 **(a)** $\frac{\sqrt{3}}{2}$ **(b)** $-\frac{\sqrt{3}}{2}$

2 **(a)** 60° and 300° **(b)** 134° and 226°

Page 36 Inequalities on graphs

1

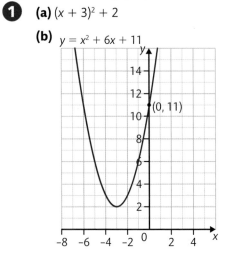

2

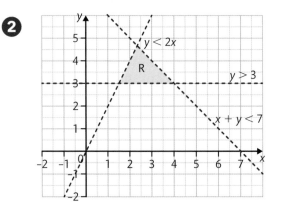

Page 37 Using quadratic graphs

(a)

x	−2	−1	0	1	2	3	4
y	6	1	−2	−3	−2	1	6

(b)

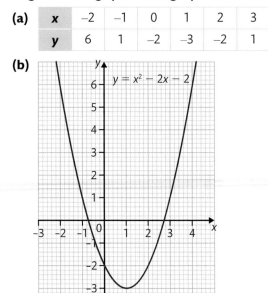

(c) **(i)** $x = 2.7$ and $x = -0.7$
(ii) $x = 3.2$ and $x = -1.2$
(iii) $x = 1.6$ and $x = -1.6$

Page 38 Turning points

1 **(a)** $2(x + 4)^2 + 1$ **(b)** $(-4, 1)$

2 **(a)** $(5, -8)$ **(b)** $(3, -51)$ **(c)** $(-4, 25)$

Page 39 Sketching graphs

1 **(a)** $(x + 3)^2 + 2$

(b) $y = x^2 + 6x + 11$

2 **(a)** $x(x-3)(x-3)$

(b)
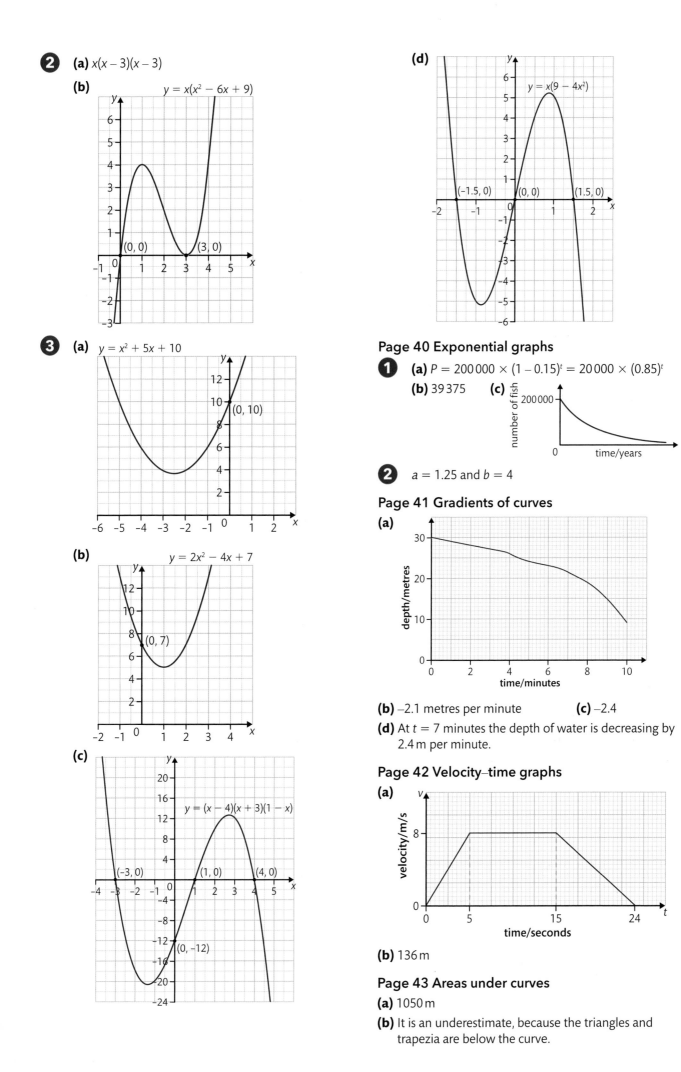
$y = x(x^2 - 6x + 9)$

(0, 0) (3, 0)

3 **(a)** $y = x^2 + 5x + 10$

(0, 10)

(b) $y = 2x^2 - 4x + 7$

(0, 7)

(c) $y = (x - 4)(x + 3)(1 - x)$

(-3, 0) (1, 0) (4, 0)

(0, -12)

(d) $y = x(9 - 4x^2)$

(-1.5, 0) (0, 0) (1.5, 0)

Page 40 Exponential graphs

1 **(a)** $P = 200\,000 \times (1 - 0.15)^t = 20\,000 \times (0.85)^t$

(b) 39 375   **(c)**

number of fish — 200 000

time/years

2 $a = 1.25$ and $b = 4$

Page 41 Gradients of curves

(a)

depth/metres

time/minutes

(b) −2.1 metres per minute   **(c)** −2.4

(d) At $t = 7$ minutes the depth of water is decreasing by 2.4 m per minute.

Page 42 Velocity–time graphs

(a)

velocity/m/s

time/seconds

(b) 136 m

Page 43 Areas under curves

(a) 1050 m

(b) It is an underestimate, because the triangles and trapezia are below the curve.

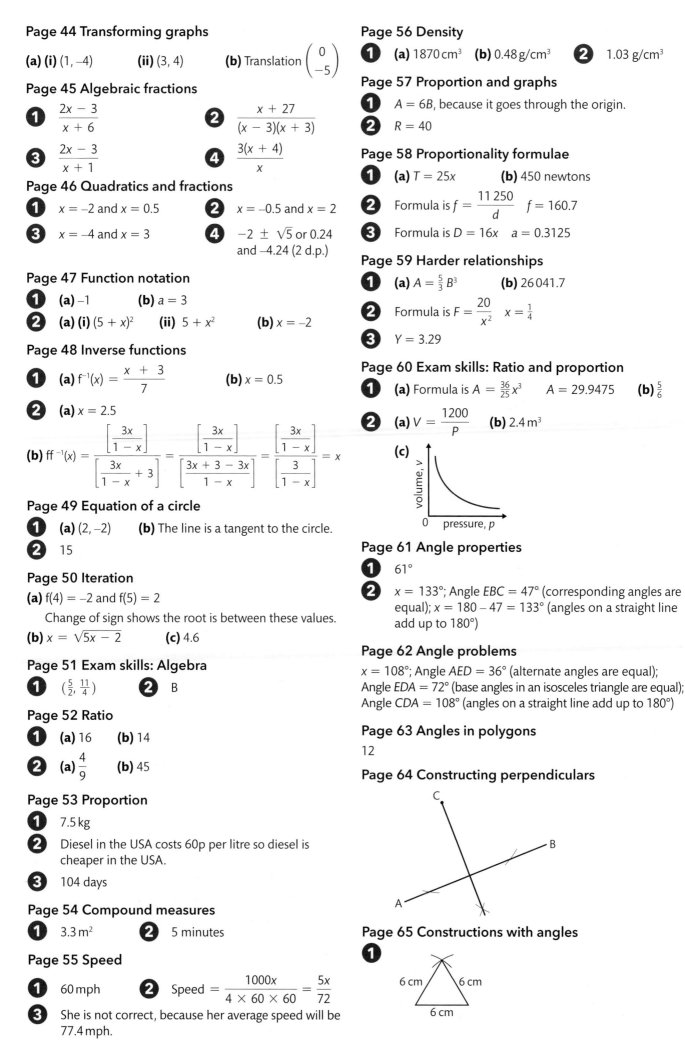

Page 44 Transforming graphs

(a) (i) $(1, -4)$ **(ii)** $(3, 4)$ **(b)** Translation $\begin{pmatrix} 0 \\ -5 \end{pmatrix}$

Page 45 Algebraic fractions

1 $\dfrac{2x - 3}{x + 6}$

2 $\dfrac{x + 27}{(x - 3)(x + 3)}$

3 $\dfrac{2x - 3}{x + 1}$

4 $\dfrac{3(x + 4)}{x}$

Page 46 Quadratics and fractions

1 $x = -2$ and $x = 0.5$

2 $x = -0.5$ and $x = 2$

3 $x = -4$ and $x = 3$

4 $-2 \pm \sqrt{5}$ or 0.24 and -4.24 (2 d.p.)

Page 47 Function notation

1 **(a)** -1 **(b)** $a = 3$

2 **(a) (i)** $(5 + x)^2$ **(ii)** $5 + x^2$ **(b)** $x = -2$

Page 48 Inverse functions

1 **(a)** $f^{-1}(x) = \dfrac{x + 3}{7}$ **(b)** $x = 0.5$

2 **(a)** $x = 2.5$

(b) $ff^{-1}(x) = \dfrac{\left[\dfrac{3x}{1 - x}\right]}{\left[\dfrac{3x}{1 - x} + 3\right]} = \dfrac{\left[\dfrac{3x}{1 - x}\right]}{\left[\dfrac{3x + 3 - 3x}{1 - x}\right]} = \dfrac{\left[\dfrac{3x}{1 - x}\right]}{\left[\dfrac{3}{1 - x}\right]} = x$

Page 49 Equation of a circle

1 **(a)** $(2, -2)$ **(b)** The line is a tangent to the circle.

2 15

Page 50 Iteration

(a) $f(4) = -2$ and $f(5) = 2$

Change of sign shows the root is between these values.

(b) $x = \sqrt{5x - 2}$ **(c)** 4.6

Page 51 Exam skills: Algebra

1 $\left(\dfrac{5}{2}, \dfrac{11}{4}\right)$ **2** B

Page 52 Ratio

1 **(a)** 16 **(b)** 14

2 **(a)** $\dfrac{4}{9}$ **(b)** 45

Page 53 Proportion

1 $7.5\,\text{kg}$

2 Diesel in the USA costs 60p per litre so diesel is cheaper in the USA.

3 104 days

Page 54 Compound measures

1 $3.3\,\text{m}^2$ **2** 5 minutes

Page 55 Speed

1 $60\,\text{mph}$ **2** Speed $= \dfrac{1000x}{4 \times 60 \times 60} = \dfrac{5x}{72}$

3 She is not correct, because her average speed will be $77.4\,\text{mph}$.

Page 56 Density

1 **(a)** $1870\,\text{cm}^3$ **(b)** $0.48\,\text{g/cm}^3$ **2** $1.03\,\text{g/cm}^3$

Page 57 Proportion and graphs

1 $A = 6B$, because it goes through the origin.

2 $R = 40$

Page 58 Proportionality formulae

1 **(a)** $T = 25x$ **(b)** 450 newtons

2 Formula is $f = \dfrac{11\,250}{d}$ $f = 160.7$

3 Formula is $D = 16x$ $a = 0.3125$

Page 59 Harder relationships

1 **(a)** $A = \dfrac{5}{3}B^3$ **(b)** $26\,041.7$

2 Formula is $F = \dfrac{20}{x^2}$ $x = \dfrac{1}{4}$

3 $Y = 3.29$

Page 60 Exam skills: Ratio and proportion

1 **(a)** Formula is $A = \dfrac{36}{25}x^3$ $A = 29.9475$ **(b)** $\dfrac{5}{6}$

2 **(a)** $V = \dfrac{1200}{P}$ **(b)** $2.4\,\text{m}^3$

(c)

Page 61 Angle properties

1 $61°$

2 $x = 133°$; Angle $EBC = 47°$ (corresponding angles are equal); $x = 180 - 47 = 133°$ (angles on a straight line add up to $180°$)

Page 62 Angle problems

$x = 108°$; Angle $AED = 36°$ (alternate angles are equal); Angle $EDA = 72°$ (base angles in an isosceles triangle are equal); Angle $CDA = 108°$ (angles on a straight line add up to $180°$)

Page 63 Angles in polygons

12

Page 64 Constructing perpendiculars

Page 65 Constructions with angles

1

2 (a) and (b)

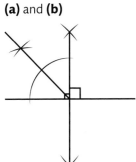

3

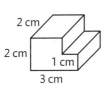
2 cm
2 cm 1 cm
3 cm

Page 66 Loci

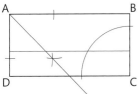

Page 67 Perimeter and area

1 1388.75 m²   **2** $x = 5, P = 40$ m

Page 68 Volumes of 3D shapes

$h = 3x$

Page 69 Surface area

144π cm²

Page 70 Prisms

1 (a) 340 cm²   (b) 400 cm³
2 (a) 720 cm²   (b) 1200 cm³

Page 71 Circles and cylinders

13.0 cm

Page 72 Circles, sectors and arcs

$\pi \times 13 \times 13 \times \dfrac{150}{360}$

Page 73 Circle facts

(a) 25°, because the angle between the tangent and the radius is 90°.

(b) 65°, because the angle in a semicircle is 90° and angles in a triangle add up to 180°.

Page 74 Circle theorems

$x = 46°$; Angle $CBA = 58°$ (alternate segment theorem; angles in a triangle add up to 180°)

Page 75 Transformations

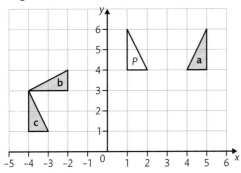

Page 76 Enlargement

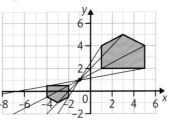

Page 77 Combining transformations

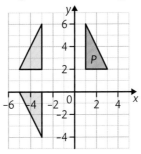

Rotation through 180° about (−1, 1)

Page 78 Bearings

1 116°

2

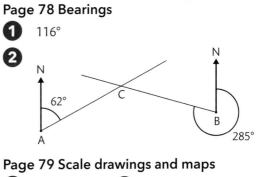
N
N
62°
C
B
A
285°

Page 79 Scale drawings and maps

1 45 m   **2** (a) 3.2 km   (b) 1 : 20 000

Page 80 Similar shapes

1 (a) 24.5 cm   (b) 2021.25 cm²   (c) 154 cm³
2 (a) Angle AEB = angle ADC and angle ABE = angle ACD (corresponding angles), so the angles in both triangles are equal.

(b) 4.3 cm

Page 81 Congruent triangles

1 $AB = CD$; $AD = BC$
BD is common in both triangles.
Therefore, triangles are congruent since SSS.

2 $AD = DB$; $AD = FE$; so $DB = FE$.
$AF = FC$; $AF = DE$; so $FC = DE$.
$BE = EC$
Therefore, triangles are congruent since SSS.

Page 82 Pythagoras' theorem

1 $11^2 \neq 5^2 + 8^2$, so it is not a right-angled triangle
2 Yes, because 53.1 m is less than 60 m

Page 83 Pythagoras' theorem in 3D

25.7 cm

Page 84 Units of length, area and volume

1 $40\,000\,\text{cm}^2$ **2** $39.25\,\text{cm}^3$ **3** $37.5\,\text{s}$

Page 85 Trigonometry: lengths

$19.2\,\text{cm}$

Page 86 Trigonometry: angles

1 $22.6°$ **2** $23.8°$

Page 87 Trigonometry techniques

1 Angle of ramp is 9°, so ramp is safe to use.

2 $4\sqrt{3}$

Page 88 Trigonometry in 3D

$18.9°$

Page 89 The sine rule

1 $15.9\,\text{cm}$ **2** $82.7°$ and $57.3°$

Page 90 The cosine rule

1 $8.61\,\text{cm}$ **2** $32.8°$

Page 91 Triangles and segments

1 $10.5\,\text{cm}$ **2** $3.53\,\text{cm}^2$

Page 92 Vectors

(a) a **(b)** $-2a$ **(c)** $b - a$ **(d)** $2(a - b)$

Page 93 Vector proof

(a) $\frac{1}{2}(p + q)$

(b) $\overrightarrow{RS} = \frac{1}{2}\overrightarrow{OP} + \frac{1}{2}\overrightarrow{PQ}$

$= \frac{1}{2}p + \frac{1}{2}(q - p)$

$= \frac{1}{2}q$

Page 94 Line segments

(a) $(6, 5)$ **(b)** $4\sqrt{5}$

Page 95 Exam skills: Geometry and measures

1 $35.3°$ **2** $32.0°$

Page 96 Probability

1 5

2 $(0.3)^3 = 0.027 < 0.1$; Sandeep is correct

Page 97 Relative frequency

1 **(a)** $\frac{24}{90} = \frac{4}{15}$ **(b)** $\frac{66}{90} = \frac{11}{15}$ **(c)** $\frac{69}{90} = \frac{23}{30}$

2 **(a) (i)** $\frac{2}{15}$ **(ii)** $\frac{7}{15}$

(b) Sample is too small so they are not accurate estimates.

Page 98 Venn diagrams

1 **(a)**

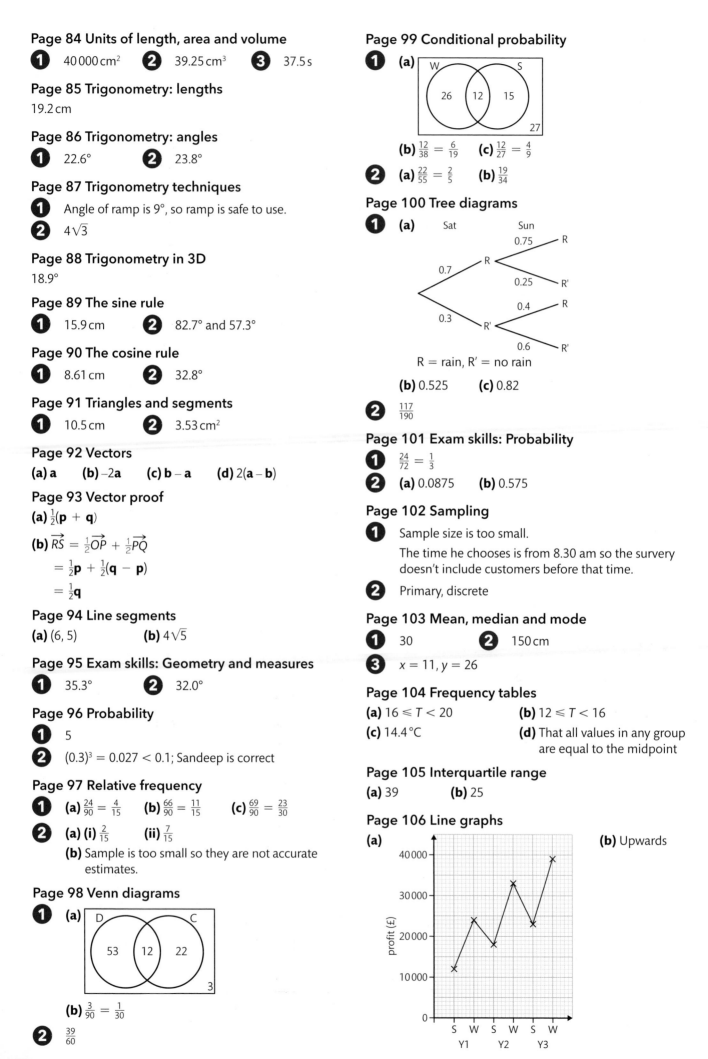

(b) $\frac{3}{90} = \frac{1}{30}$

2 $\frac{39}{60}$

Page 99 Conditional probability

1 **(a)**

(b) $\frac{12}{38} = \frac{6}{19}$ **(c)** $\frac{12}{27} = \frac{4}{9}$

2 **(a)** $\frac{22}{55} = \frac{2}{5}$ **(b)** $\frac{19}{34}$

Page 100 Tree diagrams

1 **(a)**

R = rain, R' = no rain

(b) 0.525 **(c)** 0.82

2 $\frac{117}{190}$

Page 101 Exam skills: Probability

1 $\frac{24}{72} = \frac{1}{3}$

2 **(a)** 0.0875 **(b)** 0.575

Page 102 Sampling

1 Sample size is too small.

The time he chooses is from 8.30 am so the survery doesn't include customers before that time.

2 Primary, discrete

Page 103 Mean, median and mode

1 30 **2** $150\,\text{cm}$

3 $x = 11, y = 26$

Page 104 Frequency tables

(a) $16 \leqslant T < 20$ **(b)** $12 \leqslant T < 16$

(c) $14.4\,°\text{C}$ **(d)** That all values in any group are equal to the midpoint

Page 105 Interquartile range

(a) 39 **(b)** 25

Page 106 Line graphs

(a)

(b) Upwards

Page 107 Scatter graphs

(a)

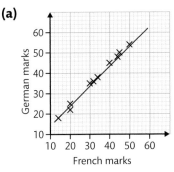

(b) As the French marks increase, the German marks increase as well.

(c) (i) 30–32   **(ii)** 37–39

Page 108 Cumulative frequency

(a) 128 g   **(b)** 11 g

Page 109 Box plots

(a) (i)

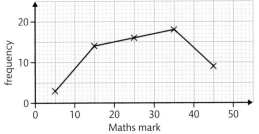

(ii)

Smallest number	3
Upper quartile	17

(b) 75

Page 110 Histograms

(a)

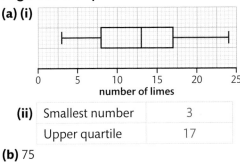

(b) Median = 42.3 minutes

Page 111 Frequency polygons

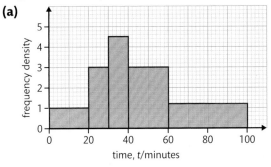

Page 112 Analysing data

The median time spent during the day is higher than the median time spent during the night, so on average people spend more time in the cafe during the day.

The range and IQR of time spent during the day are less than the range and IQR of time spent during the night.

Page 113 Exam skills: Statistics

(a)

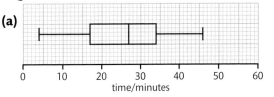

(b) The median of the girls is higher than the median of the boys so the boys complete the puzzle more quickly.

The range/IQR of the girls is higher than the range/IQR of the boys.

Page 114 Problem solving strategies

1 20   **2** $a = 2$ and $b = 3$

Page 115 Solving number problems

£12 772

Page 116 Solving proof problems

1 $\frac{1}{2}(n+1)(n+2) - \frac{1}{2}(n)(n+1)$
$= \frac{1}{2}(n^2 + 3n + 2) - \frac{1}{2}(n^2 + n)$
$= \frac{1}{2}(n^2 + 3n + 2 - n^2 - n)$
$= \frac{1}{2}(2n + 2)$
$= n + 1$

2 $(5n + 1)^2 - (5n - 1)^2$
$= 20n$
$= 4(5n)$

3 $(n + 1)^2 - (n - 1)^2 + 1$
$= 4n + 1$
$4n$ is always divisible by 2 and hence even $+1 =$ odd.

4 $n, n + 1, n + 2$
$(n + 2)^2 - n^2$
$= 4n + 4$
$= 4(n + 1)$

Page 117 Solving geometric problems

Angle $DEB = 124.2°$
The design is correct, because 124.2° is less than 125°.

Page 118 Solving algebraic problems

(a) $(34 - x)^2 + x^2 = 26^2$
$x^2 + x^2 - 68x + 1156 - 676 = 0$
$2x^2 - 68x + 480 = 0$
$x^2 - 34x + 240 = 0$

(b) Length $AB = 24$ cm

Page 119 Solving statistical problems

(a)

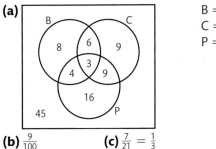

B = Biology
C = Chemistry
P = Physics

(b) $\frac{9}{100}$ **(c)** $\frac{7}{21} = \frac{1}{3}$

Published by BBC Active, an imprint of Educational Publishers LLP, part of the Pearson Education Group, 80 Strand, London, WC2R 0RL.

www.pearsonschools.co.uk/BBCBitesize
© Educational Publishers LLP 2018
BBC logo © BBC 1996. BBC and BBC Active are trademarks of the British Broadcasting Corporation.

Edited, typeset and produced by Elektra Media Ltd
Illustrated by Elektra Media Ltd
Cover design by Andrew Magee & Pearson Education Limited 2018
Cover illustration by Darren Lingard / Oxford Designers & Illustrators

The right of Navtej Marwaha to be identified as author of this work has been asserted by him in accordance with the Copyright, Designs and Patents Act 1988.

First published 2018

21 20 19 18
10 9 8 7 6 5 4 3 2 1

British Library Cataloguing in Publication Data
A catalogue record for this book is available from the British Library

ISBN 978 1 406 68609 8

Printed and bound in Slovakia by Neografia.
The Publisher's policy is to use paper manufactured from sustainable forests.

Note from the publisher
Pearson has robust editorial processes, including answer and fact checks, to ensure the accuracy of the content in this publication, and every effort is made to ensure this publication is free of errors. We are, however, only human, and occasionally errors do occur. Pearson is not liable for any misunderstandings that arise as a result of errors in this publication, but it is our priority to ensure that the content is accurate. If you spot an error, please do contact us at resourcescorrections@pearson.com so we can make sure it is corrected.

Websites
Pearson Education Limited is not responsible for the content of third-party websites.